Thermal Methods of Analysis
Principles, Applications and Problems

Thermal Methods of Analysis
Principles, Applications and Problems

P.J. Haines

With contributions by

M. Reading   T.J. Lever
F.W. Wilburn   J.W. Brown
D. Dollimore   M.R. Worland
E.L. Charsley   W. Block
S.B. Warrington

BLACKIE ACADEMIC & PROFESSIONAL
An Imprint of Chapman & Hall

London · Glasgow · Weinheim · New York · Tokyo · Melbourne · Madras

Published by
Blackie Academic and Professional, an imprint of Chapman & Hall
Wester Cleddens Road, Bishopbriggs, Glasgow G64 2NZ

Chapman & Hall, 2–6 Boundary Row, London SE1 8HN, UK

Blackie Academic & Professional, Wester Cleddens Road, Bishopbriggs, Glasgow G64 2NZ, UK

Chapman & Hall GmbH, Pappelallee 3, 69469 Weinheim, Germany

Chapman & Hall USA, One Penn Plaza, 41st Floor, New York NY 10119, USA

Chapman & Hall Japan, ITP-Japan, Kyowa Building, 3F, 2-2–1 Hirakawacho, Chiyoda-ku, Tokyo 102, Japan

DA Book (Aust.) Pty Ltd, 648 Whitehorse Road, Mitcham 3132, Victoria, Australia

Chapman & Hall India, R. Seshadri, 32 Second Main Road, CIT East, Madras 600 035, India

First edition 1995

© 1995 Chapman & Hall

Typeset in 10/12 pt Times by Photoprint, Torquay, Devon
Printed in Great Britain by The Alden Press, Oxford

ISBN   0 7514 0050 5 (PB)

A catalogue record for this book is available from the British Library

Library of Congress Catalog Card Number: 94–79172

∞ Printed on acid-free text paper, manufactured in accordance with ANSI/NISO Z39.48–1992 (Permanence of Paper)

Preface

The wide range of applications of thermal methods of analysis in measuring physical properties, studying chemical reactions and determining the thermal behaviour of samples is of interest to academics and to industry. These applications prompted the writing of this book, in the hope that the descriptions, explanations and examples given would be of help to the analyst and would stimulate the investigation of other thermal techniques.

Thermal studies are a fascinating means of examining the samples and the problems brought to us by colleagues, students and clients. If time allows, watching crystals change on a hot-stage microscope, or measuring the properties and changes on a DSC or TG or any thermal instrument can be a rewarding activity, besides providing valuable analytical information.

This book started from a series of lectures delivered at Kingston University and at meetings of the Thermal Methods Group of the United Kingdom. The collaboration and information supplied to all the contributors by colleagues and instrument manufacturers is most gratefully acknowledged, as are the valuable contributions made at meetings of the International Confederation for Thermal Analysis and Calorimetry (ICTAC) and at the European Symposia on Thermal Analysis and Calorimetry (ESTAC).

I am particularly grateful to Dr Michael Reading and Dr Fred Wilburn and to all the other contributors for lending their expertise to the production of this book. Thanks are also due to Dr Trevor Lever of TA Instruments and Dr John Barton of DRA, Farnborough for most helpful discussions. I thank the publishers for their support and advice throughout.

To Dr Fred Wilburn I should like to express the greatest personal thanks for his many helpful suggestions and contributions and for his invaluable assistance in reading the whole manuscript.

The completion of this book would not have been possible without the help, sustenance and support provided by my wife, Elizabeth. I thank her for her patience and calm during many hours of work.

Peter J. Haines

Contributors

Professor E.L. Charsley TACS, Leeds Metropolitan University, Calverley Street, Leeds LS1 3HE

Professor D. Dollimore Department of Chemistry, University of Toledo, 2801 W Bancroft Street, Toledo, OH 43606–3390, USA

Mr P.J. Haines 38 Oakland Avenue, Weybourne, Farnham, Surrey GU9 9DX

Dr T.J. Lever TA Instruments Ltd, Europe House, Bilton Centre, Cleeve Road, Leatherhead, Surrey KT22 7UQ

Dr M. Reading Physics Section, Research Department, ICI Paints, Wexham Road, Slough, Berkshire SL2 5DS

Dr. F.W. Wilburn 26 Roe Lane, Southport, Merseyside PR9 9DX

Acknowledgements

The authors wish gratefully to acknowledge permissions to reproduce figures in this book granted by the instrument manufacturers, authors and publishers listed below and identified by the references throughout the book.

American Laboratory for Figures 4.22, 4.33 and 5.12
British Antarctic Survey for Figure 6.4.6
C I Electronics Ltd for Figure 2.3
Chapman & Hall Ltd for Figure 4.20
Elsevier Science Limited for Figures 2.13, 3.44, 5.13, 5.22, 5.28 and 5.31 from *Thermochimica Acta*, and for Figure 2.18 from *Talanta*
EMAP-Maclaren for Figure 1.3
International Labmate for Figure 5.6
Laboratory Practice for Figure 1.3
Linkam Scientific Ltd for Figure 5.24a, and for help with the hot-stage microscope photographs of Figures 5.26 and 6.3.3
Mettler-Toledo for Figures 3.17, 3.28, 4.9 and 5.24b
Netzsch-Mastermix Ltd for Figures 2.5, 2.6 and 5.1
Perkin-Elmer Ltd for Figures 2.23, 3.5, 3.26, 4.14, 5.18 and 5.20
Rheometric Scientific Ltd (formerly Stanton Redcroft Ltd and Polymer Laboratories) for Figures 2.17, 2.22, 2.25, 3.40, 3.41, 4.20–4.22, 4.24, 4.30, 5.10b, 5.11 and 5.17.
The Royal Society of Chemistry for Figures 3.33 and 4.26–4.28
Dr A.J. Ryan for Figure 5.35
Setaram for Figures 3.48, 4.17 and 5.7
Shimadzu for Figures 6.21a and b
Solomat Instruments Ltd for Figures 4.31 and 4.32
TA Instruments Ltd (formerly DuPont Instruments) for Figures 2.20, 2.26, 3.4, 3.51, 3.53–3.56, 4.5, 4.12, 4.15, 4.18, 5.3, 5.10a, 6.2.7, 6.3.5 and 6.4.6
Thermal Analysis Consultancy Services, Leeds for Figures 5.2, 6.1.2 and 6.2.4
The Thermal Methods Group (UK) for Figures 3.38 and 3.39
John Wiley & Co Ltd for Figures 2.27, 2.28, 3.15, 3.24, 3.45, 5.13, 5.16, 5.29, 5.32–5.34 and 6.4.1a–e from the *Journal of Thermal Analysis*, for Figure 5.9 from *Proceedings of 2nd European Symposium on Thermal Analysis 1981*, and for Figures 2.8 and 5.30 from *Proceedings of the 7th ICTA Conference, 1982*.

Contents

Introduction to thermal methods 1

P.J. Haines

1.1
Introduction

The effects of heat on materials have fascinated and benefited humanity since the very earliest times. Even the observation of fires and the burning process was both a pleasurable, and if uncontrolled, painful experience. The use of fire to cook foods, and of ice to preserve foods, probably contributed greatly to the settlement and welfare of early peoples and cooking was perhaps the very first 'chemical experiment'. The production of both organic and inorganic pigments by heating natural materials allowed the decorative arts to develop [1].

The skills which people first acquired in the controlled use of heat allowed the manufacture of ceramics, mortars, glass and metals. Primitive apparatus dating from around 2500 BC [2] is known and the problems that arose with burning materials and the damage caused by fire are frequently seen in early settlements.

These skills, and the products they gave, were largely empirical 'arts' and their spread was jealously guarded by those who first discovered the most satisfactory technique. However, with the spread of information through travel, the methods were transmitted to people in other countries, who added their expert knowledge to improve the methods still further. The alchemists were responsible for many discoveries and their experiments in the synthesis and decomposition of natural and artificial substances laid the basis for modern chemistry [3,4]. Jabir ibn Hayyan wrote a *Book of Furnaces* and a *Book of Balances* around AD 800, but we have no evidence that he combined both [5]! In studying the history of materials, we also come to realise the effects that ageing has on their properties.

As the study of chemistry became more disciplined, the range of substances studied increased and it became necessary for scientists to be able to distinguish between different substances and materials. By studying their properties and reactions, it became possible to identify not only the constituents of a substance, but often the particular source from which it came. This is the beginning of the discipline known as *analytical chemistry*.

The modern student of chemistry or materials science may well start his or her experimental study of the subject by observing the nature of a range of materials, their appearance, mechanical properties and density and may then choose to heat the materials as a first attempt at classification [6].

A small sample, heated in a test tube, may undergo both physical and chemical changes and may alter in a large number of ways, or it may be completely stable. Table 1.1 gives some examples of behaviour that may be observed when solid substances are heated in air [6,7].

It must be noted that a single observation is often not complete in itself, but requires additional chemical or physical measurements. For example, we cannot know what gas is evolved without a simple chemical test or physical measurement, such as a spectrum. The need to use complementary analytical techniques must be recognised throughout any investigation.

The addition of some simple apparatus to determine accurately the temperature of the event, and to control the heating, or to measure colour,

Table 1.1   The effects of heat on solid materials

Effect	Possible conclusion	Example
Colour change		
Charring, burning with little residue	Organics, polymer	Paper, burning
Blackening with large residue	Metal oxide formed	$CuCO_3 = CuO + CO_2$
Metal changes to powder	Oxidation	$2Mg + O_2 = MgO$
Colour change	Transition metal salt or phase transition	HgI_2, red → yellow
Substance melts		
Melts at low temperature	Covalent?	Organics
Melts at high temperature	Ionic salts	NaCl
Substance sublimes		
White sublimate	Volatile solid Ammonium salts	NH_4Cl
Violet sublimate	Iodine	I_2
Vapours evolved (and characterised by additional tests)		
Water vapour (droplets)	Hydrates	$CuSO_4 \cdot 5H_2O$
Oxygen	Nitrates, chlorates	$2KNO_3 = 2KNO_2 + O_2$
Oxides of nitrogen (brown fumes)	Nitrates	$2AgNO_3 = 2Ag + 2NO_2 + O_2$
Carbon dioxide	Carbonates	$ZnCO_3 = ZnO + CO_2$
Physical changes		
Becomes more pliable	Plastics above T_g	
Expansion: (a) gradual	General expansion	
(b) abrupt	Phase change	
Swelling	Some intumescent materials	Polyphosphates
Shrinkage	Some strained polymers	Fibres
No effect	Stable oxides or temperature too low	MgO, Al_2O_3

plus any change in weight, or perhaps the extent of expansion or the nature of the volatiles evolved, gives a great deal more information to the analyst.

1.2 Historical development

The history of the development of thermal methods from earliest times is considered in papers by Mackenzie [2] which include an account of thermometry from the sixteenth century.

The definition of thermometric scales and the start of practical calorimetry in the eighteenth century, particularly by Lavoisier and Laplace [8] really brought about a revolution in our thinking and in the practical approach to studying the effects of heat. It led directly to the work of Fourier on heat conduction [9] and to the elegant experiments of Joule on electrical heating and calorimetry [10]. The chief drawbacks of their equipment were that it often required large samples and a long time to complete the experiment. Chemical reactions and physical measurements on gases were forerunners of today's analytical techniques

The development of scientific instruments during the earlier part of the twentieth century allowed the principles of thermal measurements to be established – for example, the measurement of the coefficients of expansion of silica by Henning [11] using an optical method led to the development of modern interferometric dilatometers [12] and the design of the thermo-balance, especially by Honda [13], has led to the modern TG systems.

In the second half of the twentieth century, vast improvements in instrumentation, sensors, data acquisition, storage and processing have been made, especially with the advent of microprocessors. The precision, sensitivity and reproducibility of modern instruments are high, and their range of temperature of operation has extended, together with the quality of temperature control.

Thermal methods of analysis are now used in a very large range of scientific investigations. Besides the more 'chemical' areas, such as polymers, fine organic chemicals and pharmaceuticals, they have applications to electronics, in construction, geology and engineering, in materials science and in quality control. They often give information impossible to obtain by other analytical methods. Very often, a complex material, such as a polymer composite, will show definite and characteristic effects on heating which relate to its nature, composition and history. These observations are informative about its properties and working life.

1.3 Definitions

Because thermal methods have been developed by many workers, it was necessary to agree on a common terminology, and the International Confederation for Thermal Analysis and Calorimetry (ICTAC) has produced definitive publications and articles, both on the nomenclature and on the calibration methods to be used [14–17]. There are still some differences in usage, but in this text we shall use the ICTAC definitions and symbols throughout.

1.3.1   *Thermal analysis*

Thermal analysis (TA) is defined [17] as:

A group of techniques in which a property of the sample is monitored against time or temperature while the temperature of the sample, in a specified atmosphere, is programmed.

The programme may involve heating or cooling at a fixed rate of temperature change, or holding the temperature constant, or any sequence of these.

The word *sample* is interpreted to mean the substance placed into the apparatus at the beginning of the experiment, and its reaction products.

The adjective is '*thermoanalytical*'.

The graphical results obtained are called the '*thermal analysis curve*', or by the specific name of the method.

The property used for study may be chosen from an extensive list, shown in part in Table 1.2.

In Table 1.2 careful distinction should be made between the terms *derivative* and *differential*. Differential techniques involve the measurement of a *difference* in the property between the sample and a reference – for example, in *differential thermal analysis (DTA)* where the difference in temperature between the sample and a reference is measured. Derivative techniques imply the measurement or calculation of the mathematical first derivative, usually with respect to time. For example *derivative thermogravimetry (DTG)* is the measurement of the *rate* of mass loss (dm/dt) plotted against temperature T.

In discussing the theories and results of thermoanalytical studies, we shall always use the 'SI' system of units and symbols [18] and, where necessary, conversion from older units will be made with the occasional exception of temperature which may be quoted in the (more familiar) degree Celsius (°C). The principal symbols and units to be used are given in Table 1.3. The usual prefixes are used, e.g. 10^{-3} m = 1 mm, 10^6 Pa= 1 MPa, 10^3 g = 1 kg, etc.

EXAMPLES OF SYMBOLS USED IN THERMAL METHODS

(a) If the symbol refers to an *object* (*the sample, the furnace*) then it should have a CAPITAL LETTER subscript:

Temperature of reference	T_R	K
Sample mass	m_S	kg
Temperature difference	ΔT	K

(b) If the symbol refers to a *phenomenon* (melting, bending) or to a *point*, then it should have a *small* letter subscript:

Table 1.2 Thermal methods

Technique	Abbreviation	Property	Uses
1. Thermogravimetry (Thermogravimetric analysis)	TG TGA	Mass	Decompositions Dehydrations Oxidation
2. Differential thermal analysis	DTA	Temperature difference	Phase changes Reactions
3. Differential scanning calorimetry	DSC	Power difference	Heat capacity Phase changes Reactions Calorimetry
4. Thermomechanical analysis	TMA	Deformations	Mechanical changes Expansions
5. Dynamic mechanical analysis	DMA	Moduli	Phase changes Polymer cure
6. Dielectric thermal analysis	DETA	Permittivity	Phase changes Polymer changes
7. Evolved gas analysis	EGA	Gases	Decompositions Catalyst and surface reactions
8. Thermoptometry		Optical	Phase changes Surface reactions Colour changes
Less frequently used techniques			
9. Thermosonimetry	TS	Sound	Mechanical and chemical changes
10. Thermomagnetometry	TM	Magnetic	Magnetic changes Curie points
11. Thermoluminescence	TL	Light emitted	Trap depths
12. Emanation thermal analysis	ETA	Gas released	Structural changes
Also used			
13. Simultaneous thermal analysis	STA	Two or more techniques used on the same sample at the same time.	
14. Controlled-rate thermal analysis	CRTA	The rate of change of the property is held constant	

Glass transition temperature	T_g	K
Melting temperature	T_m	K
Initial temperature	T_i	K
Final mass	m_f	kg

A THERMODYNAMIC DIVERSION

When a substance is heated, its physical properties and sometimes its chemical nature change, and these may be represented by chemical

Table 1.3 SI units and symbols for therrmal methods [14, 18]

Quantity	Symbol	Unit(s) and abbreviation(s)	
Basic units			
length	l	metre	m
mass	m	kilogram	kg
time	t	second	s
electric current	I	ampere	A
temperature	T	kelvin	K
amount of substance	n	mole	mol
Derived units			
energy	E	joule	$J = kg\ m^2/s^2$
power	P	watt	J/s
force	F	newton	$N(= kg\ m/s^2)$
pressure	p	pascal	$Pa\ (= N/m^2)$
concentration	c	molarity	mol/dm^3
frequency	ν	hertz	s^{-1}
Combined units			
heat	q	J	
heat capacity	C	J/K	
internal energy	U	J	
enthalpy	H	J	
free energy	G	J	
entropy	S	J/K	
thermal conductivity	k	J/(s m K)	
density	ρ	kg/m^3	
rate of reaction	v	$mol/(m^3\ s)$	
order of reaction	n		
fractional extent of reaction	α		
rate constant of nth-order reaction	k	$s^{-1}\ (mol/m^3)^{n-1}$	
activation energy	E	J/mol	
molar gas constant	R	J/(K mol)	
stress	σ	Pa	
strain	ϵ	–	
bulk modulus	K	Pa	
tensile modulus	E	Pa	
shear modulus	G	Pa	

equations and by thermodynamic functions showing the properties of the system. For a fuller treatment of the discipline of thermodynamics, the reader is advised to study a good textbook on physical chemistry, such as those listed in [19].

The *First Law of Thermodynamics* expresses the principle of conservation of energy. When heat is absorbed by a system under specified conditions – for example, at constant pressure – the heat energy of that system changes. At constant pressure this is represented by a change in the *enthalpy H*.

$$\Delta H = q_p \tag{1.1}$$

Under *standard* conditions – which are, for gases, 1 atmosphere pressure, and for liquids and solids, pure material at 1 atmosphere – we may write the *standard enthalpy* change $\Delta H^{\ominus}$, generally expressed per mole of substance. One very important value for ΔH is the *standard enthalpy of formation* $\Delta H_f^{\ominus}$ which is the standard enthalpy change when 1 mol of substance is formed from its elements in their standard states. For example:

$$Cu \text{ (pure solid)} + \frac{1}{2}O_2 \text{ (gas, 1 atm)} = CuO \text{ (pure solid)}$$
$$\Delta H_f^{\ominus}{}_{CuO, solid} = -157.3 \text{ kJ/mol}$$

The amount of heat that is absorbed by an unreacting system in raising its temperature by 1K at constant pressure is defined as the *heat capacity at constant pressure*, C_p, which is itself a function of temperature:

$$C_p = (\partial H/\partial T)_p \qquad\qquad (1.2)$$

and

$$C_p = a + bT + cT^2 \qquad\qquad (1.3)$$

Typical values of the specific heat capacity are:

Substance	C_p J/(K.g) at 20 °C
Air	1.01
Glycerine	2.43
Paraffin wax	2.09
Aluminium oxide	0.763
Aluminium	0.904

For a reacting system the difference in enthalpy between the products and the reactants under the specified conditions is realised as the heat produced by the system:

$$CaCO_3(s) = CaO(s) + CO_2(g) \qquad \Delta H^{\ominus} = 178.3 \text{ kJ/mol}$$

Heat is absorbed in this *endothermic* reaction.

$$CH_4(g) + 2O_2(g) = CO_2(g) + 2H_2O(g) \qquad \Delta H^{\ominus} = -890.4 \text{ kJ/mol}$$

Heat is evolved in this *exothermic* reaction.

The value of ΔH obtained will depend on the states of the reactants and products and on the temperature of the reaction.

Correction of ΔH to the value at a different temperature requires a knowledge of the values of C_p for all the reactants and products involved [19].

For pure substances, changes of phase from one crystal form to another, or from solid to liquid under 1 atmosphere pressure, take place at definite temperatures and with definite values of ΔH referred to as the 'latent heat'. We shall use this later in calibration of equipment.

1.3.2 Equilibrium

Most reactions tend to proceed to an *equilibrium* state. This may be almost completely reactants, or almost completely products, or any stage in between. At equilibrium, the pressures p, or concentrations c, or more strictly the *activities*, a (that is, the pressures or concentrations corrected for non-ideal behaviour) give an equilibrium constant K which is fixed for that reaction at that temperature.

$$NO_2(g) \rightleftharpoons NO(g) + \tfrac{1}{2}O_2(g)$$

$$K_p = \{(p_{NO}) \cdot (p_{O_2})^{\frac{1}{2}}/(p_{NO_2})\}_{equilibrium}$$

which has a value of about 0.04 atm$^{\frac{1}{2}}$ at 500 K.

Where substances are in their standard states, we may consider their activities as 1, so that, for the calcium carbonate reaction:

$$CaCO_3(s) \rightleftharpoons CaO(s) + CO_2(g)$$

$$K = \{a_{CaO} \cdot a_{CO_2}/a_{CaCO_3}\}$$

reduces to

$$K_p = \{p_{CO_2}\}_{equilibrium}$$

The *Second Law of Thermodynamics* deals with the tendency of reactions to proceed. The driving force of a reaction depends on both the enthalpy change ΔH and the *entropy change* ΔS. The entropy S of a system is related to W, the number of different molecular energy level arrangements of the system, where k is the Boltzmann constant:

$$S = k \ln W$$

or, rather loosely, to the *randomness* of the system. Thus, gases have high entropies; regular crystalline solids low entropies.

The *Third Law of Thermodynamics* states that the entropies of perfect crystalline solids at 0 Kelvin may be assigned a value of zero.

These thermodynamic quantities are combined to give the *free energy G*:

$$G = H - TS$$

For a process at constant temperature:

$$\Delta G = \Delta H - T \Delta S$$

For a reaction to proceed spontaneously, the free energy must decrease, so the value of ΔG must be negative.

For a reaction at equilibrium, ΔG is zero; but $\Delta G^{\ominus}$ is related to the equilibrium constant by the reaction isotherm equation:

$$\Delta G^{\ominus} = -RT \ln K$$

and the variation of K with temperature can be shown to be

$$\ln K = -\Delta H^{\ominus}/RT + \Delta S^{\ominus}/R$$

known as the van't Hoff equation, or reaction isochore. Similar equations may be applied to physical equilibria, such as solution, adsorption or vaporisation.

Chemical reactions occur at particular rates, defined by the rate of change of amount of substance with respect to time. Often it is more convenient to use a quantity related to the amount, such as pressure, concentration, or, in the case of solids, the *fraction reacted* α.

The rate of reaction is defined as dn/dt and is related to the amounts of reactants by a rate law:

$$dn/dt = k \cdot f(n)$$

where k is the rate constant, and $f(n)$ is a function of the amount, n. For example, when we consider the random degradation of a polymer, we may write the rate equation as:

$$d\alpha/dt = k \cdot (1-\alpha)$$

Other reactions need much more complicated functions of n or α, which will be discussed in Chapter 2.

The rate of reaction varies with temperature, chiefly due to the Arrhenius condition that species must possess energy in excess of the *activation energy*, E, if the reaction is to be effective. This gives rise to the Arrhenius equation:

$$k = A \cdot \exp\left(- E/RT\right)$$

where A is referred to as the 'pre-exponential factor' and depends on the orientation and structure of the reactants.

For most reactions involved in thermal methods, the rate laws are complex, but it is important to realise that the rate of change will depend very greatly on two factors:

(a) the amount of substance reacting and
(b) the temperature.

Anything which affects these, such as the heat evolved by an exothermic reaction or the loss of reactant by vaporisation, will change the rate. It is worth noting that a *derivative* curve, such as a DTG curve, indicates the rate of reaction.

1.3.3 *General apparatus*

Almost all thermal methods use a similar arrangement of apparatus, as shown in Figure 1.1.

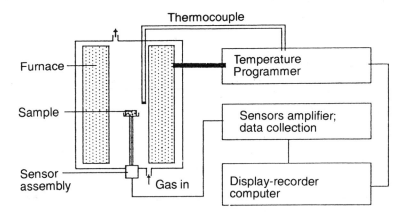

Figure 1.1 Schematic diagram of thermal analysis system.

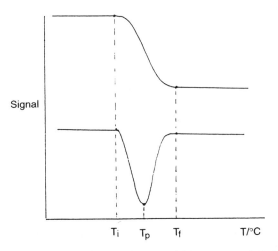

Figure 1.2 Typical thermal analysis curves showing initial, peak and final temperatures.

The *sample* is often put into a container or crucible which is then placed in contact with the *sensor* which is to measure the particular property. A temperature sensor must also be present at (or very near) the sample to follow its temperature as the experiment progresses.

The sensor assembly then fits into a special *furnace* and the *atmosphere* around the sample is established. This is most important, because we may wish to study (or to avoid!) reactions with air or other reactive gases.

The furnace is controlled by a temperature programmer and is set by the operator to raise (or lower) the temperature of the furnace at whatever rate has been chosen as most suitable. The data are collected by the sensor system and, after processing, are displayed on a screen or recorder as a *thermal analysis curve*.

If you go into any modern laboratory, there are often as many computers as there are instruments. The computer forms a vital part of the 'armoury' of any analytical laboratory, and its use in combination with thermal analysis techniques is very important.

The use of computers may be divided into approximately four parts:

1. As a means of instruction in the operation of the instrument.
2. To make the collection, interpretation, storage and retrieval of instrumental data easier for the operator.
3. To allow the user more easily, and more accurately, to calculate the results of an experiment.
4. To simulate the behaviour of the instrument, or of the sample, under special conditions.

The general interconnections of a computer with a thermal analysis system is shown in Figure 1.3.

One or more instruments are connected to the computer via the experimental control interface and also via the data acquisition interface. The user interacts with the computer via the keyboard and the VDU screen. The data produced may be stored in a large variety of ways, but most systems involve both hard disc and floppy disc storage. Display of results, hard copy of thermal analysis curves and of calculated parameters, such as coefficients of expansion from a TMA experiment, may be produced on printers or plotters.

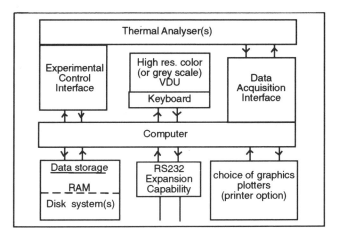

Figure 1.3 Schematic of a typical computer system. (From [20] with permission.)

Computers may be used in the planning of the experimental work by writing a method into the computer system. The method can preset the analyser to the selected starting conditions, load the sample from a robotic autosampler into the analyser and carry out the run, controlling the heating programme and the purge gases. The data obtained may then be analysed according to a selected regime and a report presented to the analyst with the thermal analysis traces, the results for, say, melting point and heat of fusion from a DSC trace, plus an assessment of the results on each sample for purposes of quality control [21].

1.4.1   *The computer as an instructional tool*

There are several 'computer-aided learning' schemes available for analytical chemistry, notably from the ACOL organisation [22]. Similar systems may be written for most commercial instruments, and are often available from the manufacturers. As an example, the 'PL-Show' computer programs [23] give both an overview, and also specific applications for a wide range of thermal analysis equipment.

The program can instruct, as a 'bench textbook' in the definitions, mode of operation and applications of a selected thermal method. A menu system leads the user through the choice of method and conditions, such as sample size and heating rate, and gives specific instructions for the operation of the particular instrument involved. Many such systems are 'fail-safe'. For example, they will not allow you to continue if you choose the wrong temperature limits (which might burn out the furnace), or if you have not turned on the analyser, or even if there is an obstruction that would impede the movement of the furnace!

1.4.2   *The computer as a data gatherer*

The earliest commercial thermal analysis instruments produced signals that were transmitted to a chart recorder. This had the advantage of simplicity, and presented an immediate, visual reminder of the 'state of the experiment'. It was, however, impossible to retain the signal that was produced, and so to process or do direct calculations on it. With the coming of mechanical and then electronic integrators, the analyst was no longer forced to measure peaks with a planimeter, or cut them out and weigh them!

The modern computer-linked thermal analysis system will do much more than replace a recorder. A large memory allows for storage of both the data and the applications programs. Special software for particular techniques is provided. While a run is in progress, data from a previous run may be analysed.

1.4.3 *The computer as a data processor*

The software available with computer-linked thermal analysis instruments allows the rapid, accurate determination of results from the experimental runs. Most of the examples and problems presented in this book have been analysed by computer. The integration of peaks in DTA and DSC, the determination of the glass transition temperature from DSC or TMA and the accurate measurement of mass loss in TG may all be most easily carried out using the computer software. Calculations of any type are much quicker, and probably more accurate, for the average operator when carried out by the computer. Because of this, it is becoming more important to consider, and report, the computer specifications that were used. What type of A/D conversion was used and how many bits were involved? How much smoothing was used and what algorithm was employed? How many signal points were taken and how were they treated and stored [24]?

While modern instruments should produce excellent data with a low signal/noise ratio, the sample does not always do what is expected! Often complex, overlapping and sloping traces make the reasoning unclear. Computer users often quote the acronym 'GIGO': 'Garbage In – Garbage Out.' In some cases, it is possible that the user has not thought sufficiently about the experimental conditions, or the sample preparation. In such a case, however good the computer or the software, the result is likely to be wrong!

1.4.4 *The computer as a research tool*

The design and operation of thermal analysis equipment, even more so than other analytical instruments, depends very much on the conditions used. Obviously, a change in the original amount of sample will alter the change in mass observed in a TG experiment, but it may also alter the type of chemical reaction that will occur. The effects of sample size, heating rate, the thermal conductivity of the sample and its container may all be simulated using suitable computer programs. Attempts to discover the most appropriate kinetic equation to describe the chemical reactions occurring may be tried using the computer. Side-by-side comparison of runs obtained from samples of different origin may reveal subtle differences.

**1.5
Factors affecting
thermal analysis results**

Many analytical methods give a result that is specific to the compound under investigation. For example, the infrared spectrum of polystyrene is characteristic of that material, and depends little on the sampling method, or instrument, or the time taken to perform the run. Many thermal

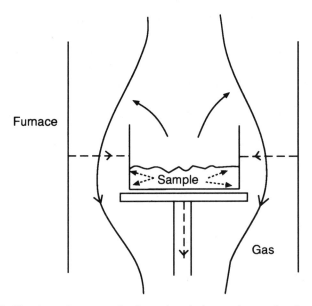

Figure 1.4   The dynamic nature of a thermal analysis experiment, showing the flow of heat (dashed lines) and of gases (solid lines), plus the progress of sample reaction.

methods are much less 'compound specific' and the results obtained may depend upon the conditions used during the experiment. The reasons for this will become apparent in later chapters of this book, but may briefly be summarised as due to the *dynamic* nature of the processes involved (Figure 1.4). The signal generated by the sensors will depend on the extent and rate of reaction, or the extent and rate of change of the property measured. The transfer of heat by conduction, convection and by radiation around the apparatus, and the interaction of the surroundings with the sample all affect these changes.

It is therefore essential, whenever a thermal analysis experiment is reported, that the precise conditions used are included in the report. Similarly, comparison of samples should only be made when their curves are run under the same conditions, or when differences in conditions are clearly stated.

The main factors which must be considered and reported for every thermal analysis experiment are given below.

1.5.1   *The sample*

The chemical description of the sample should be given, together with its source and pre-treatments. It is also important to state its purity, chemical composition and formula, if these are known. If the sample is diluted with another substance, such as an inert reference material, then the composition of the mixture should be given. This diluent can greatly alter the thermal conductivity of the sample. Even the particle size may alter the shape of the curve, especially where a surface reaction is involved.

For example: hydrated copper sulphate, $CuSO_4 \cdot 5H_2O$, powder 50–60 μm, XYZ Co. Ltd, GPR grade, diluted 50% in alumina.

All of these factors could affect the thermal analysis curve. A change in the chemical composition would alter the decomposition of a copper sulphate sample, as would a change in purity. The diluent will alter the heat transfer characteristics and the shape of the curve.

The 'history' of the sample, and possibly its method of preparation, can affect the curve. Trace amounts of impurity can catalyse decompositions, or cause different reaction sequences.

1.5.2   *The crucible*

The material of the sample holder or crucible should be stated, together with a guide to its geometry. Sample holders are chosen because they will not interact with the sample during the course of the experiment. However, changing from aluminium to platinum, or to silica or alumina, may well alter the heat transfer because of the different thermal conductivity, and possibly the chemistry of the process – for example, if a reaction occurs that could be catalysed by platinum and we are using a platinum crucible.

The geometric shape of the sample holder is also important, since a shallow, wide holder will allow free diffusion of reactant gases to the sample and the diffusion of product gases away, while a narrow, deep holder may restrict gas flow.

1.5.3   *The rate of heating*

Experiments may be carried out at heating rates (dT/dt) from 0 K/min – that is, under isothermal conditions – through the 'normal' rates of 10 K/min, to very high rates over 100 K/min, which are sometimes used to simulate burning of materials. Similarly, experiments may be conducted with different *cooling* rates Since the rates of heat transfer, of physical change and of most reactions are finite, the sample will react differently at different heating rates.

The transfer of heat between the source, such as the furnace, and the different parts of the sample or reference materials is *not* instantaneous, but will depend on the conduction, convection and radiation that can occur within the apparatus. There is bound to be a *thermal lag* between different parts of the apparatus, and the higher the rate of heating, the greater this lag is likely to be. General equations describing the corrections to be applied when working at different heating rates, β, are often of the form [25]:

$$T_e = T_{e,0} + a \cdot \beta + c$$

where T_e is the observed, uncorrected temperature at heating rate β, $T_{e,0}$ is the corrected temperature at $\beta = 0$, a is a constant, β is the heating rate (in K/min) and, c is a correction determined by calibration.

For example, for gallium melting:

$$T_e = 303.56 + 0.074 \cdot \beta - 0.2 = 304.10 \text{ K}$$

A rapid heating rate may allow some of the sample to melt before it decomposes, while at a slow heating rate all reaction has occurred below the melting temperature. In order to approach equilibrium conditions more closely, we should use a *very small* heating rate, preferably less than 1 K/min. The heating rate will affect the start, range and finish of a change:

$$T_{i,fast} \quad > \quad T_{i,\,slow}$$
$$T_{f,\,fast} \quad > \quad T_{f,\,slow}$$
$$(T_f - T_i)_{fast} \quad > \quad (T_f - T_i)_{slow}$$

The resolution of two changes which occur at temperatures close to each other is likely to be much better at slow heating rates than at fast.

Heating rates need not be linear, and may follow a complex pattern which must be specified in detail: isothermal at 25 °C for 2 min, followed by 50 K/min up to 300 °C; isothermal at 300 °C for 10 min, then cool at 5 K/min to room temperature.

1.5.4   *The atmosphere*

The transfer of heat and the chemistry of the sample reaction will depend greatly upon the atmosphere surrounding the sample and its products.

Even if there is no reaction between the sample and the atmosphere, the heat transfer by the gas will affect the results. Consider four gases commonly used in thermal methods:

Gas at 1 atm	*Thermal conductivity at* 373 K/(10^{-2} J/(smK))
Helium	17.77
Nitrogen	3.09
Air	3.17
Carbon dioxide	2.23

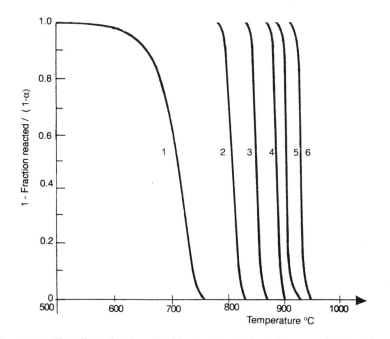

Figure 1.5 The effect of carbon dioxide pressure on the TG curves of AR calcium carbonate. 100 mg heated at 1 °C/min in (1) 1.0 atm N_2, (2) 0.1 atm CO_2, (3) 0.3 atm CO_2, (4) 0.5 atm CO_2, (5) 0.7 atm CO_2, (6) 1.0 atm CO_2. In runs (2)–(5), N_2 was mixed with CO_2 to give a total pressure of 1 atmosphere. (From [29] with permission.)

The heat transfer by conduction through helium will be nearly eight times greater than through carbon dioxide.

When reaction occurs, the nature of the thermal event will change completely. While a metal may be stable in helium, it could oxidise in air. When a chemical equilibrium exists, for example, between calcium carbonate, calcium oxide and carbon dioxide:

$$CaCO_3 \rightleftharpoons CaO + CO_2$$

then Le Chatelier's Principle tells us that the presence of a large concentration of product will force the equilibrium back towards the reactants. Thus, although calcium carbonate may start to decompose below 700 °C in air containing little CO_2, in CO_2 at 1 atmosphere pressure the decomposition does not occur until over 900 °C [26-28]. Figure 1.5 shows this effect.

Sometimes the sample produces its own 'self-generated atmosphere'. This can affect the equilibrium, the kinetics and the heat transfer of the experiment.

We should also consider the *flow rate* of the gas. A static atmosphere will not sweep away reaction products from the sample, or act as a heat transfer agent.

1.5.5 *The mass of the sample*

The physical properties, size and the packing or density will alter the results obtained. If very small samples are used, perhaps less than 1 μg, then the signal produced will be small, and in certain cases each crystal may react at a different time producing multiple peaks. Very large samples will produce a much greater effect, but if several changes occur they may not be fully resolved. When we wish to analyse mixtures in which small amounts of one component need to be detected – for example, a coating on a filler used in a plastic formulation – then large samples may be necessary. Comparison of thermal analysis curves should be done only when comparable sample masses are used.

These parameters are important for any thermal analysis experiment, and for particular methods other details may also be needed. As an aid to remembering the five most important parameters, perhaps the acronym 'SCRAM' may be helpful:

Sample
Crucible, or sample holder
Rate of heating
Atmosphere
Mass of sample.

1.6 Simultaneous and complementary techniques

A single thermal method does not always give sufficient information to allow the analyst to be sure what is occurring. For example, a 'downward' peak produced by a DTA experiment means that an endothermic event occurs over this temperature range. It does not tell you whether this is a chemical reaction or a physical change such as melting, or whether any gases are evolved. A TG experiment on the same sample may show a mass loss over this temperature range, thereby ruling out melting, but still not identifying any volatiles. Conversely, if the TG curve showed no mass loss, the process of melting could be confirmed by direct observation of the sample. It is clear that combining several analytical methods gives a better profile of the changes taking place, although we may have to sacrifice the optimum conditions for each separate technique.

If the two techniques are performed on different samples, or at very different times, then the methods are referred to as *complementary*. If a DTA experiment is done on one sample of a polymer, and a TG experiment is done on a second sample of the same polymer, the techniques are complementary.

Such a scheme might also involve a thermogravimetric experiment from which the solid product is then removed and used to produce an X-ray diffraction pattern.

If two techniques are performed on a single sample at the same time, then they are *simultaneous* techniques. Since a hyphen is often used to separate the abbreviations, for example TG-DTA, they are also called 'hyphenated techniques'. An apparatus which allows a combination of differential scanning calorimetry and simultaneous sample observation using a microscope would be DSC-thermoptometry [17].

Wherever possible, simultaneous techniques are to be preferred, since different samples and different conditions may alter the thermal analysis curves, as we have noted above. However, optimum conditions may be very different for the two methods. For example, the highest sensitivity of a DTA experiment is attained at high heating rates while, for TG, the best resolution is attained at low heating rates!

Very nearly every analytical method has been employed as a complementary technique to a thermal method. Gases evolved during a reaction have been separated by gas chromatography, dissolved and titrated, measured electrochemically and spectroscopically, particularly by infrared and by mass spectrometry. Solid products have been analysed for their surface area and catalytic activity and by every sort of compositional analysis method. It must be stressed as firmly as possible that these additional analyses and experiments are *vital* to a full understanding of the results obtained by thermal analysis.

**1.7
Problems**
(solutions on p. 273)

1. A small sample of a white powder is provided, and you decide to examine the effects of heat on it. On gentle heating over boiling water, it melts. On stronger heating in air, on a spatula, it burns, leaving little residue. What conclusions do you make about the nature of this substance, and what simple technique or thermal methods might you use to study it further? How could you prevent it burning in the apparatus?

2. A second sample is provided. It is a pink, crystalline solid. On gentle heating it changes to blue and gives off a vapour which condenses to droplets on the cooler part of the tube. On stronger heating it gives brown fumes and leaves a black solid. What conclusions can you draw from these observations?

3. A thermal analyst performs an experiment in which the dimensions of a sample are measured as it is heated. Suggest a name for this technique and decide whether it is a special case of one of the major thermal methods.

4. Complete the following definition, by comparison with the standard definition of thermal analysis:

Thermoelectrometry is a family of techniques in which an . . . . . . property of the sample is measured against . . . . . . or temperature

while the . . . . . of the . . . . . in a specified . . . . . is programmed.

Are any of the techniques in Table 1.2 members of this family?

5. Can you spot the error(s) or omissions in the following experimental details? 'A sample of a complex compound was heated in a thermal analysis apparatus, in a platinum sample holder, up to 500 °C and the thermal analysis curve recorded.'

6. What complementary or other techniques could be used to study the changes and reactions in questions 1 and 2 above?

References

1. G.I. Rochlin (ed.), *Scientific Technology and Social Change*, Scientific American, New York, 1977.
2. R.C. Mackenzie, *Thermochim. Acta*, 1984, **73**, 249.
3. J. Read, *Prelude to Chemistry*, Bell, London, 1939.
4. W.H. Brock, *The Fontana History of Chemistry*, Fontana Press, London, 1992.
5. E.J. Holmyard, *Alchemy*, Pelican Books, London, 1957.
6. *Vogel's Textbook of Qualitative Inorganic Analysis* (5th edn; revised, G. Svehla), Longman, London, 1979.
7. H.T. Clarke, *A Handbook of Organic Analysis* (5th edn; revised, B. Haynes), Arnold, London, 1975.
8. A.L. Lavoisier, P.S. de Laplace, *Mem. R. Acad. Sci.*, Paris, 1784, 355.
9. J.B. Fourier, *Theorie Analytique de la Chaleur*, Paris, 1822.
10. J.P. Joule, *Collected Works*, The Physical Society, London, 1884.
11. F. Henning, *Ann. Physik*, 1907, **23**, 809.
12. R. Kato *et al.*, *Thermochim. Acta*, 1988, **134**, 383.
13. K. Honda, *Sci. Rep. Tohoku Univ.*, 1915, **4**, 97.
14. R.C. Mackenzie *et al.*, *Talanta*, 1969, **16**, 1227; 1972, **19**, 1079.
15. R.C. Mackenzie *et al.*, *J. Thermal Anal.*, 1975, **8**, 197; *Thermochim. Acta*, 1981, **46**, 333.
16. R.C. Mackenzie in *Treatise on Analytical Chemistry*, P.J. Elving (ed.), Pt I, Vol. 12, Wiley, New York, 1983.
17. J.O. Hill, *For Better Thermal Analysis and Calorimetry III*, ICTA, 1991.
18. (a) IUPAC Manual of Symbols, *Pure Appl. Chem.*, 1979, **51**, 1; 1982, **56**, 1239. (b) I. Mills, *Quantities, Units and Symbols in Physical Chemistry*, IUPAC, Blackwell, Oxford, 1988.
19. (a) P.W. Atkins, *Physical Chemistry*, OUP, Oxford, 1978; (b) G.M. Barrow, *Physical Chemistry* (5th edn), McGraw-Hill, Singapore, 1988.
20. J.A. Hider, *Lab. Practice*, 1986, Feb., 19.
21. B. Wunderlich, *Int. Lab.*, 1982, 32.
22. *Analytical Chemistry by Open Learning* (ACOL), Wiley, Chichester, England.
23. Rheometric Scientific Ltd, Surrey Business Park, Epsom, Surrey.
24. J.G. Dunn, *J. Thermal Anal.*, 1993, **40**, 1435.
25. E.L. Charsley *et al.*, *J. Thermal Anal.*, 1993, **40**, 1409.
26. E.L. Simons, A.E. Newkirk, *Talanta*, 1964, **11**, 549.
27. F.W. Wilburn, J.H. Sharp, D.M. Tinsley, R.M. McIntosh, *J. Thermal Anal.*, 1991, 37, 2003.
28. R.M. McIntosh, J.H. Sharp, F.W. Wilburn, *Thermochim. Acta*, 1991, **165**, 281.
29. F.W. Wilburn, J.H. Sharp, *J. Thermal Anal.*, 1993, **40**, 133.

General bibliography

Note Books which deal specifically with ONE thermal method will be listed at the end of the most relevant chapter. All the books listed here discuss a range of thermal methods.

A. Blazek, *Thermal Analysis*, Van Nostrand, London, 1972.
M.E. Brown, *Introduction to Thermal Analysis*, Chapman & Hall, London, 1988.

E.L. Charsley, S.B. Warrington (eds), *Thermal Analysis – Techniques and Applications*, RSC, Cambridge, 1992.

T. Daniels, *Thermal Analysis*, Kogan Page, London, 1973.

J.W. Dodd, K.H. Tonge, *Thermal Methods*, ACOL/Wiley, London, 1987.

J.L. Ford, P. Timmins, *Pharmaceutical Thermal Analysis*, Ellis Horwood, Chichester, 1989.

P.D. Garn, *Thermoanalytical Methods of Investigation*, Academic Press, New York, 1965.

G. van der Plaats, *The Practice of Thermal Analysis*, Mettler, 1991.

D.N. Todor, *Thermal Analysis of Minerals*, Abacus Press, Tunbridge Wells, 1976.

E.A. Turi, *Thermal Characterisation of Polymeric Materials*, Academic Press, New York, 1981.

W.W. Wendlandt, *Thermal Analysis* (3rd edn), Wiley, New York, 1986.

G. Widman, R. Reisen, *Thermal Analysis: Terms, Methods, Applications*, Huthig, Heidelberg, 1987.

B. Wunderlich, *Thermal Analysis*, Academic Press, New York, 1990.

CHAPTERS IN GENERAL ANALYTICAL TREATISES AND TEXTBOOKS

G.W. Ewing, *Instrumental Methods of Chemical Analysis*, Wiley, New York, 1969.

F.W. Fifield, D. Kealey, *Principles and Practice of Analytical Chemistry* (3rd edn), Blackie, Glasgow, 1990.

I.M. Kolthoff, P.J. Elving, C.B. Murphy (eds), *Treatise on Analytical Chemistry*, Part 1: *Theory and Practice* (2nd edn), Vol. 12, Section J, Wiley, New York, 1983.

D.A. Skoog, D.M. West, *Principles of Instrumental Analysis*, Holt-Saunders, Tokyo, 1980.

C.L. Wilson, D.W. Wilson (eds), *Comprehensive Analytical Chemistry*, Vols XII, A–D, Elsevier, Amsterdam, 1981–4.

JOURNALS IN ENGLISH

Thermochimica Acta, Elsevier, Amsterdam.
Journal of Thermal Analysis, Wiley, Chichester.
Journal of Chemical Thermodynamics, Academic Press, New York.

Very many papers in chemical, polymer, mineral and instrumental journals.

Abstracts were collected in *Thermal Analysis Abstracts*, which became *Thermal Analysis Reviews and Abstracts* (TARANDA), but which has now ceased publication.

2 Thermogravimetry

P.J. Haines

2.1
Introduction The measurement of the mass of a sample is often one of the earliest quantitative experiments we do. Every day, many portions of foods, pharmaceuticals, containers, chemicals, polymers or engineering parts are weighed and recorded. The extension of this simple technique into analysis provides the discipline of *gravimetric analysis*. This has been defined [1] as 'quantitative analysis by weight' and is the process of isolating and weighing an element or a definite compound of the element in as pure a form as possible. Gravimetric experiments to determine water of crystallisation, and to study oxidation and reduction, are used to teach the principles of stoichiometry, formulae and analysis. Many gravimetric techniques involve a reaction to produce a definite compound and then heating it to convert to a stable material of known composition. For example, bismuth may be precipitated as a cupferron complex, but must then be ignited 'strongly' to give the stable oxide Bi_2O_3 [1].

The samples that the analyst is given need to be studied in the form 'as received' and frequently this neither is a definite compound nor is it pure! While we shall try to perform quantitative analysis, the answers will often be very dependent on the treatment regime and the method of analysis. As a practical example, the moisture content of a sample of soil may be measured by drying below 100 °C and its content of organic materials by igniting at around 500 °C [2].

If we can follow all the effects of heat on the changing mass of the sample, we should be able to extract information on the stages of analysis and on the temperatures needed to bring each stage to completion.

Measurement of *mass* on a balance is done by the process of weighing. That is, a balance compares the unknown mass with a standard mass by comparing their *weights*, which have the units of *force* (newtons) and the magnitude of *mass times gravitational acceleration*. The mass of a body stays *constant*, but the weight will alter if the acceleration acting on the body alters.

2.2
Historical The origins of thermogravimetry have been fully documented by Duval [3], Keattch and Dollimore [4] and Wendlandt [5]. The most significant

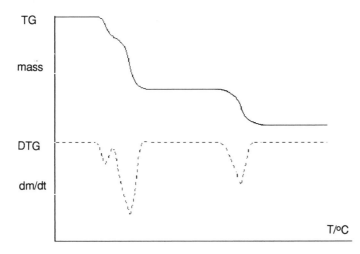

Figure 2.1 Typical TG (solid) and DTG (dashed) curves.

contribution was made by Honda [6] in 1915 who used a lever-arm balance fitted with an electrical furnace to investigate manganese oxysalts. Simple experimental apparatus has been described which uses modified manual analytical balances and controlled electric furnaces [7]. The Chevenard thermobalance was the first to record weight changes automatically using a photographic method. This was used extensively from 1936 by Duval and others [8].

The development of the electronic microbalance [9] allowed smaller samples and furnaces to be used, and work to be carried out in controlled atmospheres and in vacuum.

**2.3
Definition of
thermogravimetry**

Thermogravimetry (TG) [10] is a technique in which the *mass* of the sample is monitored against time or temperature while the temperature of the sample, in a specified atmosphere, is programmed.

It should be recognised that several manufacturers and users prefer to call this technique *thermogravimetric analysis* (TGA). This avoids confusion with the glass transition temperature, T_g. The apparatus is called a *thermobalance*, or less frequently, a thermogravimetric analyser.

In order to enhance the steps in the thermogravimetric curve, the *derivative thermogravimetric* (DTG) trace is frequently drawn. Remember that this is the plot of the *rate of mass change, with time*, dm/dt.

**2.4
Apparatus**

As with most thermal analysis systems, the thermobalance has four major parts:

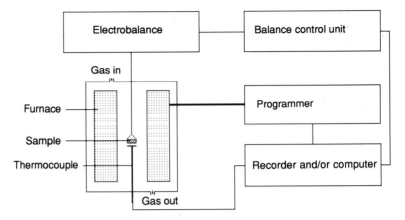

Figure 2.2   Schematic diagram of thermobalance system.

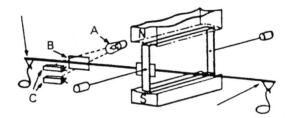

Figure 2.3   Schematic of a microbalance: A, lamp; B, shutter; C, photocells.
(Courtesy CI Electronics Ltd.)

- the electrobalance and its controller
- the furnace and temperature sensors
- the programmer or computer
- the recorder, plotter or data acquisition device.

2.4.1 *The balance*

The balance used in many of the commercial apparati is a modified electronic microbalance. One example is shown in Figure 2.3. The light-weight arm is pivoted about an electrical coil suspended in a magnetic field. The position of the arm is measured by an optical sensor and any deflection

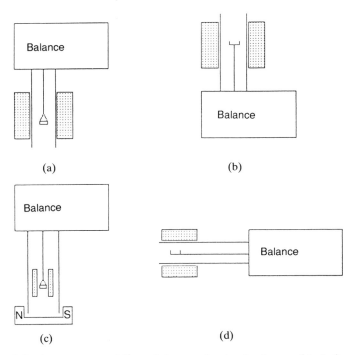

Figure 2.4 Arrangements of thermobalances, showing the furnace (shaded), sample position and casing: (a) and (c) suspended; (b) top-loading; (d) horizontal. The magnet position for Curie point calibration is shown in (c). (After [11].)

causes a current to be supplied to the coil which restores it to the 'null' position. This is important for maintaining a constant position of the sample in the same zone of the furnace. The sample suspended into the furnace from one arm of the balance is counterbalanced by a tare, either electrically applied to the coil or added to the reference pan. The balances will often give a resolution of 1 μg or better, and sample sizes range from a few milligrams up to 30 g. With this very high sensitivity, freedom from vibration is essential.

Several different arrangements of the balance and furnace are possible, as shown in Figure 2.4 [11], and the geometry may affect the results obtained.

The difficulties that may affect the operation of a balance used in conjunction with a furnace are most important. The presence of corrosive or oxidising gases near the balance mechanism is most undesirable, and the balance enclosure is often *purged* with an inert gas such as nitrogen at room temperature. Problems with static electricity, which can cause the attraction of the sample pan onto the glass enclosure, may be overcome by coating the surfaces with a conducting film or use of antistatic spray.

The movement of the gas surrounding the sample due to convection may cause noise and affect the observed mass. This may be reduced by introducing baffles within the sample enclosure. Usually experiments are conducted in a flowing atmosphere, and suitable design reduces convective effects.

Thermomolecular flow may occur when the balance is operated at low pressure [12]. If the sample or its suspension or the counterpoise experience temperature gradients when the pressure is low and the mean free path is comparable to the dimensions of the apparatus, the flow of gas molecules causes an apparent mass change.

With a temperature difference of 100 K, and a pressure of 10^{-1} torr, the effect can be as much as 100 μg.

When samples of large volume must be used, there is an effect due to Archimedes' Principle that 'when a body is immersed in a fluid it experiences an upthrust equal in magnitude to the weight of fluid displaced' [13]. If the volume of sample plus sample holder is V m^3 and the pressure p Pa, then the *mass* of an ideal gas of molar mass M displaced at temperature T will be:

$$m = MpV/RT$$

As the sample is heated, so the density of the gas decreases, while the density of the solids hardly changes. Therefore the upthrust will decrease, and the sample will *apparently gain* in measured mass.

As an example, consider a sample plus crucible of volume 1 cm^3, that is 10^{-6} m^3, at atmospheric pressure, approximately 10^5 Pa of nitrogen gas ($M = 0.028$ kg). If the molar gas constant R is 8.314 J/(K mol) then the buoyancy will change by about 0.8 mg between 300 K and 1000 K, so there will be an apparent weight *gain* of 0.8 mg.

There are several ways to compensate for this effect. Firstly, we may run a 'blank' buoyancy curve with an inert sample of similar volume and subtract this curve from the experimental curve for our sample.

Secondly, the use of very small samples and crucibles will reduce the effect. If the volume is reduced to 2.8×10^{-3} cm^3, equivalent to 10 mg of MgO, then the apparent gain in mass is only 2.2 μg.

Thirdly, if we use a twin furnace system as shown in Figure 2.5 [14,15] where a counterpoise of similar volume is heated at the same time, the effects largely cancel.

Thermogravimetry in very corrosive atmospheres may be carried out using a magnetically suspended balance [16,17], as shown in Figure 2.6. An electromagnet suspended from the microbalance holds a glass-encased permanent magnet in the measuring cell. The sample in the furnace is connected by a suspension wire to the permanent magnet and each mass change is transmitted back without contact. This allows the use of very reactive atmospheres, or of samples which produce corrosive or toxic products, without their coming into contact with the balance mechanism.

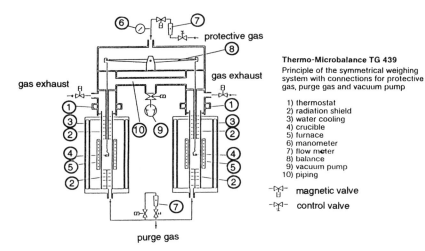

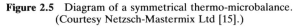

Thermo-Microbalance TG 439

Principle of the symmetrical weighing
system with connections for protective
gas, purge gas and vacuum pump

1) thermostat
2) radiation shield
3) water cooling
4) crucible
5) furnace
6) manometer
7) flow meter
8) balance
9) vacuum pump
10) piping

magnetic valve

control valve

Figure 2.5   Diagram of a symmetrical thermo-microbalance.
(Courtesy Netzsch-Mastermix Ltd [15].)

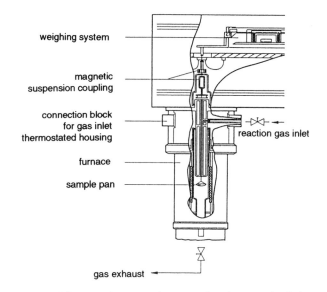

Figure 2.6   Diagram of a magnetic suspension thermo-microbalance.
(Courtesy Netzsch-Mastermix Ltd [17].)

2.4.2 *Furnace*

The furnaces used are generally non-inductively wound electrical resistance heaters, although infrared and microwave heating have been suggested. The furnace and balance arrangements are shown in Figure 2.4. The most important features needed from the furnace are listed below:

1. The furnace should have a zone of uniform temperature which is considerably longer than the sample plus holder. This is much easier to achieve with small samples and holders.
2. The heat from the furnace must not affect the balance mechanism. Often baffles are placed between the two to reduce transfer of heat.
3. The furnace should be capable of rapid response and a range of heating and cooling rates, and should be capable of heating to temperatures well above those of interest. Rapid cooling is also most useful when a quick 'turn round' time is needed when many samples must be run.
4. The furnace lining should be inert at all temperatures used. A tube of ceramic such as alumina or mullite, or, for lower temperatures, silica glass may surround the sample, or in some cases, furnace and sample.

2.4.3 *Programmer*

The furnace temperature is measured by a thermocouple, generally platinum–platinum 13% rhodium, which is most suitable up to 1600 °C, is chemically fairly inert but has a fairly low output of about 8–12 μV/K. Another thermocouple often used is chromel–alumel which is suitable to 1100 °C, has a larger response of about 40 μV/K, but is more easily oxidised.

The signal from the furnace (or control) thermocouple is transmitted to the programmer and the temperature it represents is compared with that required by the programme set by the operator. If the temperature is too low, the system must respond by supplying more power to the furnace, and, if too high, by reducing the power. The response times of the controller and of the furnace govern the thermal lag of the instrument and hence the range of heating rates that is achievable. A slow response programmer controlling a very large furnace may only allow heating rates below 10 K/min, whereas a small mass furnace and a fast controller will allow rates up to 100 K/min [18].

2.4.4. *Samples*

Thermogravimetry is most frequently carried out on solids. The sample should be obtained in an approved way [19,20] so that it provides a meaningful analysis. This is easy with pure, homogeneous compounds, but very difficult with large samples which may differ from one region to another. Procedures such as 'coning and quartering' of a finely ground sample may give better reproducibility, but often the very differences that we observe between samples are significant. One region of a sample may be more porous, or more easily oxidised, or have a higher moisture content than the bulk of the sample. Crystalline samples may behave differently from fine powders, especially where surface reactions are involved. It is also very often difficult to produce a fine powder from samples such as thermoplastics.

Samples should ideally be small, powdered and spread evenly in the crucible. Typical thermogravimetric runs might be done under the conditions given below.

Sample:	Calcium oxalate monohydrate
Crucible:	Platinum pan
Rate:	10 K/min
Atmosphere:	Nitrogen, 20 cm^3/min
Mass:	10.5 mg

Other examples are given later in the book.

2.4.5 *Temperature calibration*

In order that the balance shall operate with as little interference as possible, the sample temperature is often measured by a thermocouple placed near to, but *not in contact with* the sample.

Sample temperature thermocouples are generally of the same types used for furnaces, but frequently rather smaller and more delicate, since they are nearer the sample and the products of reactions.

The temperature measured by a thermocouple which is not in contact with the sample must be subject to a thermal lag which could be several degrees. Unless the experiment is carried out isothermally on a very small sample, this lag will vary with temperature and with the reactions proceeding, the furnace atmosphere, the heating rate and the geometry of the system. It is very important to calibrate the thermobalance in conditions which reproduce those which will be used in actual experiments, and to be able to reproduce the conditions on a variety of instruments. A survey was carried out by ICTAC of the most appropriate method of temperature calibration of thermobalances and concluded that a sharp, reproducible physical change would give the best results. The materials selected are metals and alloys which are ferromagnetic at low temperature, but which lose their ferromagnetism at a well-defined Curie point [21].

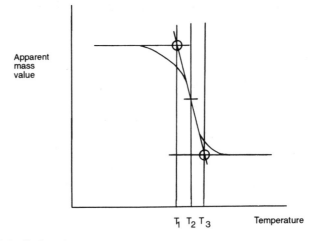

Figure 2.7   Curie point curve using the arrangement of Figure 2.4(c). For nickel, the reported values are: $T_1 = 351.4\ °C$; $T_2 = 352.8\ °C$; $T_3 = 354.4\ °C$. T_2 is taken as the Curie point for calibration.

Table 2.1   Curie point calibration samples for TG

Sample	$T_{Ci}/(°C)$
Permanorm 3	259
Nickel	353
Mumetal	381
Permanorm 5	454
Trafoperm	750

If the magnetic material is placed in the sample holder and a magnet placed below the sample, as shown in Figure 2.4(c), the magnetic force on the sample will cause an apparent increase in mass. In a thermogravimetric run in nitrogen, the force will disappear fairly abruptly at the Curie point, as shown in Figure 2.7. The mid-temperature, or the peak of the DTG curve, is now used as a fixed point for calibration using the NIST materials listed in Table 2.1 [11,22].

Another method of calibration uses wires made from the high purity materials of very accurately known melting points such as those used in calibrating DTA [see Chapter 3]. These wires are used to attach a mass to the sample pan, which will detach immediately the wire melts [23].

A rough check on the calibration may be performed with known reactions, such as those given as examples later in this chapter, and ensuring that the DTG peak temperatures match those obtained on the apparatus when it has been recently calibrated by a standard method.

2.4.6 *Atmosphere*

The need to surround the sample in an inert or a reactive atmosphere, and to control the evolution of gases from the sample, generally means that thermogravimetry is conducted in a flowing gas stream. Too high a flow rate can disturb the balance mechanism, while too low a flow rate will not remove product gases or supply reactant gas. Therefore, a flow rate of about 10–30 cm^3/min is often used. The flow of gas can also contribute to the transfer of heat and assist the transfer of products to any external gas analysis system [see Chapter 5].

If we wish to investigate a reaction that produces gas:

$$A(s) \quad = \quad B(s) \; + \; C(g)$$
$$\text{e.g.} \quad CaCO_3 \quad = \quad CaO \; + \; CO_2$$

then we can conduct the experiment in a flowing inert atmosphere such as nitrogen, where the carbon dioxide will be swept away as soon as it is formed, or in an atmosphere of carbon dioxide, which will shift the position of equilibrium, according to Le Chatelier's Principle. If we consider a simple thermodynamic approach to the reaction, we have shown in Chapter 1 that the equilibrium constant for the reaction is given by:

$$K_p = \{p_{CO_2}\} \text{ equilibrium}$$

Since the reaction is *endothermic*, K_p increases as the temperature increases according to the van't Hoff equation:

$$\ln (K_p) = - \Delta H^{\ominus}/RT + \text{constant}$$

so the equilibrium pressure of CO_2 becomes greater.

At a low partial pressure of CO_2, decomposition will take place at a lower temperature than when the partial pressure is high.

If the sample is confined, then the atmosphere around it is made by the sample itself, that is, it is *self-generated*. This will cause similar effects to an externally imposed high pressure of product gases [24,25].

If we want the best contact between sample and surrounding gas, we must spread the sample thinly, or use a porous or gauze container, taking care that the sample holder is inert to all samples and gases [26,27].

**2.5
Kinetics of reactions**

The study of reactions can be divided into three parts:

1. The stages, intermediates and products of reaction.
2. The energetics of the reaction stages.
3. The reaction mechanism and reaction kinetics.

The first two parts may readily be studied by TG and DSC, and the third may also be investigated by thermal methods.

Consider a typical endothermic solid-state reaction:

$$A \text{ (solid)} = B \text{ (solid)} + C \text{ (gas)}$$

As this reaction proceeds, gas and mass will be lost and heat absorbed. Several equations are used to model the process, but they may not be valid in all situations!

1. The extent of reaction ξ is a quantity having the dimension amount of substance and defined by:

$$n_B = n_{B,0} + \nu_B \cdot \xi$$

where n_B is the amount of substance B, $n_{B,0}$ is a chosen amount of B, e.g. at $t = 0$, and ν_B is the stoichiometric number of B, positive if B is a product, negative if B is a reactant [28].

For reactions in solution the progress of reaction is usually followed by the changes in concentration c_B of the species B with time, but for solid-state reactions this is hardly appropriate and it is more usual to follow the changes in the *fraction reacted*, α, with time.

The *rate of reaction* may now be written in terms of α:

$$\text{Rate} = d\alpha/dt$$

The rate of a reaction, even if measured at constant temperature, usually varies with time and with the value of α. It is therefore common to write:

$$\text{Rate} = d\alpha/dt = k_T, f(\alpha) \tag{2.1}$$

where k_T is the rate constant at temperature T, and $f(\alpha)$ is a mathematical expression in α.

It would be nice if the same mathematical $f(\alpha)$ applied throughout the reaction, as is quite often the case for solution and radiochemical reactions, but it sometimes alters to a different $f'(\alpha)$ part-way through the reaction, due to the changes in mechanism, geometry or chemistry. It is easy to see why this should be so if we consider a thermogravimetric experiment where the different reaction stages overlap!

If a reaction is studied simultaneously by thermogravimetry and by a group-specific technique such as IR, the values of α may not be the same, since the IR-active species may not contribute to the greatest mass loss!

Relations used for the kinetics of solid-state reactions

Sestak and Berggren [29] summarised the many equations relating the rate of solid-state reactions to α. They may be written as a combined equation:

$$d\alpha/dt = k_T \cdot \alpha^m \cdot (1-\alpha)^n \cdot (-\ln(1-\alpha))^p$$

Table 2.2   Kinetic equations

Type	$f(\alpha)$ = Rate/k	$g(\alpha)=kt$
Order equations		
F1 First order	$(1-\alpha)$	$-\ln(1-\alpha)$
F2 Second order	$(1-\alpha)^2$	$1/(1-\alpha)$
Geometric		
R2 Contracting area	$2(1-\alpha)^{1/2}$	$1-(1-\alpha)^{1/2}$
R3 Contracting volume	$3(1-\alpha)^{2/3}$	$1-(1-\alpha)^{1/3}$
Acceleratory		
P1 Power law $(m>1)$	$m(\alpha)^{(m-1)/m}$	$\alpha^{1/m}$
Sigmoid curves		
An Avrami–Erofe'ev $(n=2, 3$ or $4)$	$n(1-\alpha)\cdot$ $(-\ln(1-\alpha))^{(n-1)/n}$	$[-\ln(1-\alpha)]^{1/n}$
B1 Prout–Tompkins	$\alpha(1-\alpha)$	$\ln(\alpha/(1-\alpha)) + c$
Diffusion		
D1 1-D diffusion	$1/(2\alpha)$	α^2
D2 2-D diffusion	$[-\ln(1-\alpha)]^{-1}$	$[(1-\alpha)\ln(1-\alpha)] + \alpha$
D3 3-D diffusion	$(3/2)(1-\alpha)^{2/3}$ $[1-(1-\alpha)^{1/3}]^{-1}$	$[1-(1-\alpha)^{1/3}]^2$
D4 Ginstling–Brounshtein	$(3/2[(1-\alpha)^{1/3}-1]^{-1}$	$[1-2\alpha/3]-(1-\alpha)^{2/3}$

It is sufficient to quote three examples and refer to the more comprehensive list in Table 2.2.

(a) One-dimensional diffusion-controlled reaction:

$$d\alpha/dt = k_T/2\alpha$$

(b) Two-dimensional growth of nuclei (Avrami equation):

$$d\alpha/dt = k_T(1-\alpha)\cdot(-\ln(1-\alpha))^{1/2}$$

(c)   First-order reaction: random decay of active species:

$$d\alpha/dt = k_T(1-\alpha)$$

Upon integration, this last equation gives:

$$\ln(1/(1-\alpha)) = k_T\cdot t$$

or, for a general kinetic equation:

$$g(\alpha) = k_T\cdot t$$

The interpretation of the kinetic equation considers the way in which the reaction starts, by a process of nucleation, then how those nuclei grow and what reaction or interface geometry is involved, and finally, how the reactants decay [30].

2. The rate constant, k, depends greatly on temperature and is thus sometimes written k_T. The equation used for many reactions is that of Arrhenius:

$$k_T = A \exp(-E/RT) \tag{2.2}$$

where A is the pre-exponential factor; E is the activation energy (J/mol) and R is the molar gas constant (8.314 J/(K mol)).

Several thermal analysis papers lament that different values of E have been found by different workers [31]. This may be due to a change in reaction mechanism during reaction, to a true variation in E or to differences caused by experimental procedure. The validity of the Arrhenius equation is very often assumed, and deviation attributed to changes in the mechanism [32].

3. The temperature in a non-isothermal experiment is ideally controlled, often to fit a linear programmed rise, β K/min:

$$T_t = T_0 + \beta \cdot t \tag{2.3}$$

This again is an ideal situation and for large sample masses, highly exothermic or endothermic reactions may cause changes in the real sample temperature. Urbanovici and Segal [33] suggest a modified equation:

$$T_t = T_0 + \beta \cdot t + s(t) \tag{2.3a}$$

where $s(t)$ is the difference between the sample temperature and the programmed temperature.

Therefore, it is as well when starting to study the kinetics of a reaction to beware of these pitfalls. A good guide to strategy for studying kinetics is given by Galwey [34] who advises the use of isothermal methods as well as rising temperature methods.

It may be as well to stress that the scheme recommends using complementary techniques, such as hot-stage and scanning electron microscopy, to confirm the mechanism of reaction.

Problems arise when equations (2.1)–(2.3) are combined and manipulated mathematically.

2.5.1 *Measurement of α and dα/dt*

For a thermogravimetric curve showing a single step, such as Figure 2.9, we may calculate α at a particular time, or temperature from the measured mass m_t and the initial and final masses, m_i and m_f.

$$\alpha = (m_i - m_t) / (m_i - m_f) = m_i / (m_i - m_f) - m_t / (m_i - m_f)$$

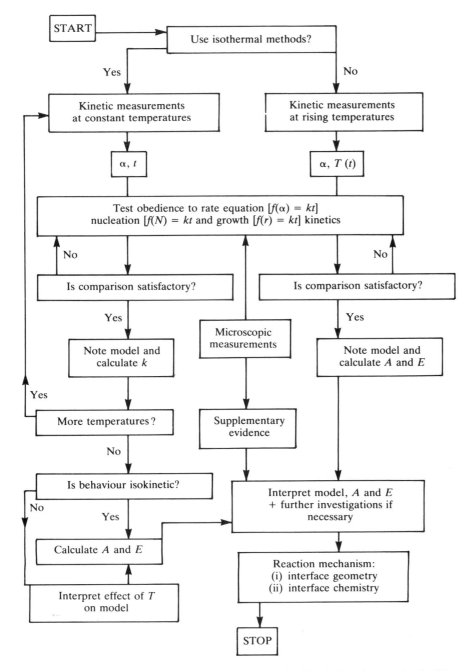

Figure 2.8 Schematic representation of the steps involved in the kinetic analysis of solid-state reactions. (From [34] with permission.)

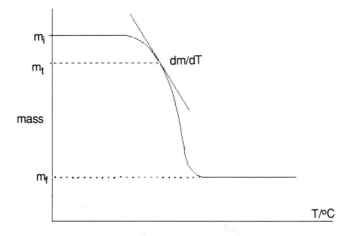

Figure 2.9   Schematic TG curve and the analysis for kinetics.

Differentiating this gives:

$$d\alpha/dt = -[dm_t/dt] \, / \, (m_i - m_f)$$

and we may therefore measure the rate of reaction from the slope of the mass–time curve. Since the DTG curve already measures dm_t/dt, this may be used to find $d\alpha/dt$ directly.

If we combine the linear heating equation (equation (2.3)), then:

$$dT/dt = \beta \tag{2.4}$$

and $d\alpha/dt = (d\alpha/dT) \cdot (dT/dt) = (d\alpha/dT) \cdot \beta$.

For a DSC curve, if the total area A represents the total reaction over the whole time or temperature interval and the fractional area α gives the partial reaction up to time t, then

$$\alpha = a/A$$

It has also been pointed out that the ordinate deflection y_t in DSC represents the rate of heat supply in J/s, so that we might also use it to find the rate, since $d\alpha/dt$ is proportional to y_t.

DIFFERENTIAL EQUATIONS

From equations (2.1) and (2.2):

$$d\alpha/dt = A \cdot \exp \, (-E/RT) \cdot f(\alpha)$$

Combining with equation (2.4)

$$d\alpha/dT = (A/\beta) \cdot \exp \, (-E/RT) \cdot f(\alpha) \tag{2.5}$$

Table 2.3

$T(K)$	$10^3k/T$	α	$(1-\alpha)$	$\ln(1-\alpha)$	$10^5.d\alpha/dT$	$\ln(d\alpha/dT)$
929	1.076	0.005	0.995	−0.005	6.25	−9.680
938	1.066	0.011	0.989	−0.011	8.86	−9.332
966	1.035	0.047	0.953	−0.048	19.28	−8.554
994	1.006	0.118	0.882	−0.125	37.51	−7.888
1013	0.987	0.194	0.806	−0.215	52.10	−7.560
1032	0.969	0.297	0.703	−0.277	68.77	−7.282
1050	0.944	0.432	0.568	−0.566	86.48	−7.053
1069	0.935	0.598	0.402	−0.911	102.6	−6.882
1078	0.928	0.691	0.309	−1.175	105.2	−6.857

Rearranging the variables and taking logarithms:

$$\ln(d\alpha/dT) - \ln(f(\alpha)) = \ln(A/\beta) - E/RT \qquad (2.6)$$

By plotting the left side of equation (2.6) against $1/T$, the activation energy can be obtained from the slope, and A from the intercept. In order to do this, we must have some idea of which $f(\alpha)$ we should use from Table 2.2!

Example For the decomposition of calcium carbonate, the kinetic equation could be one of many possible, for example:

1. Diffusion (D2: 2-D diffusion):

$$d\alpha/dt = k \cdot (- \ln(1-\alpha))^{-1}$$

2. Prout–Tomkins (B1: interacting nuclei):

$$d\alpha/dt = k \cdot \alpha \cdot (1-\alpha)$$

3. Mampel (R3: contracting sphere):

$$d\alpha/dt = 3k \cdot (1-\alpha)^{2/3}$$

We can test these by plotting the appropriate data according to equation (2.6). For example, the R3 plot would be of $\{\ln(d\alpha/dT) - (2/3)\cdot \ln(1-\alpha)\}$ against $1/T$.

For this experiment 10 mg of $CaCO_3$ powder was heated at 10 °C/min in a platinum crucible in flowing nitrogen. The TG plot was measured and selected data are given in Table 2.3 above.

It is very clear that the most appropriate kinetic equation is R3, the contracting sphere or Mampel equation. This agrees well with published data [35,36] although some references suggest an R2 law. The slope of the Mampel line is $-E/R$, and gives a value of E of about 190 kJ/mol. This is a little high, but comparable to some published data. Reading has pointed out [31] that the range of values quoted for the activation energy of this reaction is very large indeed! The pre-exponential factor A is about 10^7 s^{-1}.

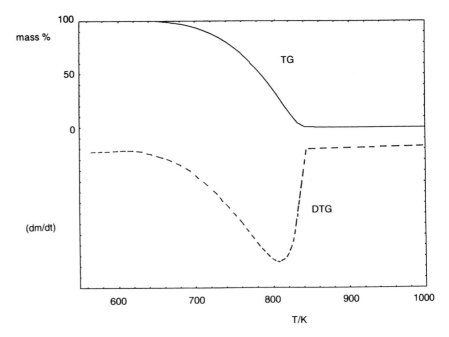

Figure 2.10   TG and DTG curves for 10 mg of $CaCO_3$ powder heated at 10 K/min in flowing nitrogen (30 cm³/min).

REDUCED TIME PLOTS

In order to check the validity of the kinetic equation and which $f(\alpha)$ should be used, it is good practice to plot all results onto the same graph [4,37]. For isothermal cases, plotting the value of α against $(t/t_{0.5})$ often gives a single trace of characteristic shape, as shown in Figure 2.12. If the value of $f(\alpha)$ is evaluated from the equations of Table 2.2, the fit of the observed data can identify the kinetic rate equation. For example, for a first-order (F1) mechanism,

$$\alpha = 1 - \exp(-0.693(t/t_{0.5}))$$

Similar reduced plots have been suggested for non-isothermal results [38] and for constant rate results [39].

INTEGRAL METHODS

If we attempt to use the integrated rate equations represented by the functions $g(\alpha)$ in Table 2.2, the integration of equation (2.5) is required.

$$g(\alpha) = \int d\alpha/f(\alpha) = (A/\beta) \cdot \int (\exp(-E/RT)) \, dT \qquad (2.7)$$

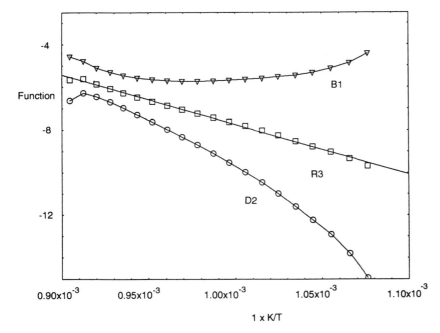

Figure 2.11 Plots of $\ln(d\alpha/dT) - \ln(f(\alpha))$ against $1/T$ for diffusion (D2), Prout-Tomkins (B1) and Mampel (R3) kinetics applied to calcium carbonate decomposition.

This involves the difficult integration of the right-hand side temperature integral. If we assume that $\alpha = 0$ until the reaction starts at T_i, we may change the limits of integration and integrate from 0 to T:

$$\int (\exp(-E/RT))\ dT = (E/R) \cdot \int [(\exp(-x))/x^2]\ dx = (E/R) \cdot p(x)$$

where $x = E/RT$ and $p(x) = [(\exp(-x))/x^2]$.

Approximations to determine $p(x)$ have been proposed, one of the best of which is

$$p(x) \approx \exp(-x)/[x(x+2)]$$

Values of the function $p(x)$ may be calculated by computer and have been tabulated. Various approximations have been suggested for the temperature integral, for example: Doyle [40];

$$\log_{10} p(x) \approx -2.315 - 0.4567x \quad \text{(for } x > 20\text{)} \tag{2.8}$$

Coats and Redfern [41]:

$$g(\alpha) = (ART^2/\beta E)\ [1-(2RT/E)] \cdot \exp(-E/RT)$$

Brown [42] gives an excellent discussion of the mathematics used and the treatment of kinetic results. Ozawa [43] and Flynn and Wall [44] have

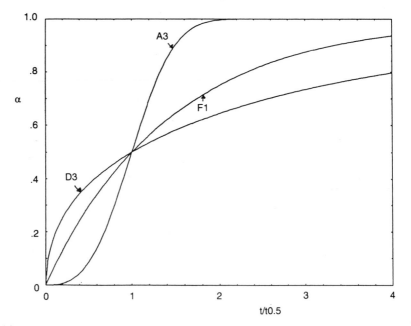

Figure 2.12   Reduced time plots for Avrami (A3), first-order (F1) and 3-D diffusion (D3) kinetics.

shown that if different heating rates, β, are used, the temperature T_α for a given conversion α may be shown, by combining equations (2.7) and (2.8), to be

$$\log_{10} \beta = -2.315 + \log_{10} (AE/R) - \log_{10} g(\alpha) - 0.4567(E/RT_\alpha) \quad (2.9)$$

Plotting $\log \alpha$ against $1/T_\alpha$ generally gives good straight lines of slope $-0.4567\ (E/R)$. The pre-exponential factor, A, is also obtained by equation (2.10)

$$A = \beta \cdot (E/RT^2) \cdot \exp(E/RT) \quad (2.10)$$

A revision of the value of E determined by this method uses a correction value of $D = -\ \mathrm{d} \ln(p(x))/\mathrm{d}x$, which is tabulated in ASTM E 698, or may be evaluated by computer.

Example   Since the kinetics of decomposition of polymeric materials is a very important study and relates to their stability and performance, the thermal decomposition of poly(tetrafluoroethylene), PTFE, was studied by TG [45] using samples of about 10 mg, and flowing air at 30 cm^3/min.

Rate, β(K/min)	T(K) (2%)	T(K) (10%)
2.5	775	798
5.0	781	805
10	791	818
20	802	829

The slopes of the plots are -1.80×10^4 and -1.67×10^4 K, so that

$E \approx -(1/0.4567) \times R \times \text{slope} \approx -2.19 \times R \times \text{slope}$
$E \approx 328$ kJ/mol and 304 kJ/mol,   giving an average
$E = 316$ kJ/mol

The values are refined by calculating $x = E/RT$ for a temperature in the middle of the range, say 790–810 K, and looking up D in tables. For an average value:

$x = 47.5$,   so that $D = 1.042$
$E \approx -2.303 \times R \times \text{slope}/D = 319$ kJ/mol

This procedure may be repeated until a constant value of E is obtained. Using the mid-point value of 10 K/min

$A = 10 \ (319\ 000/(8.314 \times 800^2)) \cdot \exp(319\ 000/(8.314 \times 800))$
$A = 4.05 \times 10^{20}$ min^{-1}

It may be noted that, for reactions having several stages, the value of E may change as a function of the degree of conversion, and this has been illustrated with the pyrolysis of poly(1,4-phenylene-di(methoxyphenyl)-vinylene) in argon [46] where the reaction shows three stages with activation energies of 200, 263 and 258 kJ/mol.

CONSTANT RATE METHODS

If we consider that the rate of reaction ($d\alpha/dt$) is controlled by the value of the temperature, the fraction reacted and the nature of the atmosphere, then control of these parameters could be used to control the rate directly.

Rouquerol [47] proposed that 'a quantity directly related to the decomposition rate is kept constant'. This quantity may be the evolved gas flow, the thermal flow or the DTG signal. This is the technique of constant rate thermal analysis (CRTA). The quasi-isothermal, quasi-isobaric techniques of the Pauliks [48] controlled their system in such a way as to keep the rate of mass loss constant, and to confine the reaction gases so that they were lost at constant atmospheric pressure.

Reading [39] has described the 'rate jump' technique, where the temperature and pressure are controlled such that, at a temperature T_1, the rate of mass loss is constant at $(d\alpha/dt)_1$, and when it is rapidly raised to a new temperature T_2, the new constant rate is $(d\alpha/dt)_2$. For a small interval, the value of $\ln[f(\alpha)]$ changes little during the jump, so that:

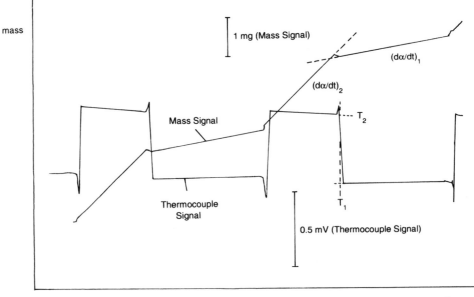

Figure 2.13   Typical results for rate jump constant rate thermal analysis [39].

$$E = \frac{-R \ln[(d\alpha/dt)_1/(d\alpha/dt)_2]}{(T_2 - T_1)/(T_1 \cdot T_2)}$$

For calcium carbonate, a value of E= 193 kJ/mol was reported.

2.6
Applications of thermogravimetry

A selection of examples is given below, but the range of applications is very large indeed. Readers are advised to study the specialist texts on thermogravimetry for further particulars.

Note!   It must be remembered that TG will ONLY detect changes where the mass of the sample changes, by reaction, or vaporisation, or because a physical property changes its interaction with an imposed external force, e.g. the magnetic permeability in an imposed magnetic field.

2.6.1   *Thermogravimetric curves*

DECOMPOSITION OF MAGNESIUM HYDROXIDE $(Mg(OH)_2)$

This is found as a natural mineral, brucite; it is used in gravimetric analysis and as a fire-retardant additive in polyamides, polyolefins and polyesters.

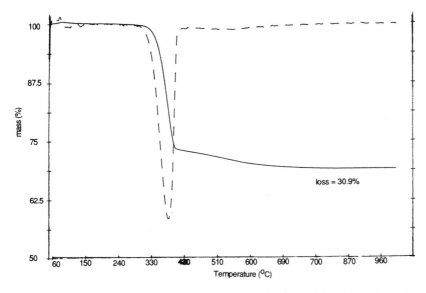

Figure 2.14 TG and DTG curves for magnesium hydroxide, 7.04 mg, heated in a Pt crucible at 10 K/min, nitrogen.

The thermogravimetric trace of a commercial sample given in Figure 2.14 shows a single mass loss over a fairly broad temperature range. The sample is stable up to about 200 °C, and then loses about 27% of its mass rapidly by 450 °C; a much slower loss then occurs up to 800 °C with a total mass loss of 30.9%.

The chemical equation for the reaction is:

$$Mg(OH)_2(s) = MgO(s) + H_2O(g)$$

and the water evolved can be detected by chemical, physical or spectroscopic techniques (see Chapter 5). The stoichiometry of the above reaction suggests that the calculated loss should be

$$\% \text{ Loss} = 100 \times M(H_2O)/ M(Mg(OH)_2) = 100 \times 18.0/58.3$$
$$= 30.87\%$$

Thus we see a very good agreement between theory and experiment in this case. The reason for the two stages shown on the TG and DTG curves is less obvious. It has been attributed to the rapid reaction to decompose all the hydroxide and to lose the majority of the water vapour, followed by the gradual slow diffusion of water vapour adsorbed on, or trapped within, the oxide [49]. Another alternative is the presence of a small amount of carbonate impurity [50].

Samples prepared by different reactions, or stored under different

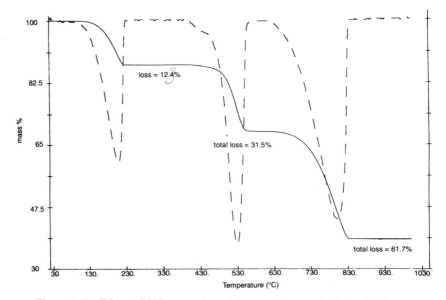

Figure 2.15   TG and DTG curves for calcium oxalate monohydrate, 12.85 mg, Pt crucible, 20 K/min, nitrogen.

conditions, or subjected to different packing [51] have all been shown to give different TG curves, although the quantitative nature of the change is unaffected.

The endothermic nature of the dehydration reaction is one reason for the use of magnesium hydroxide as a fire retardant, and it is found that incorporating it into a polymer changes the decomposition of *both* the polymer and the hydroxide.

CALCIUM OXALATE MONOHYDRATE ($CaC_2O_4 \cdot H_2O$)

The oxalate hydrates of the alkaline earth metals, calcium, strontium and barium, are all insoluble and have been used in gravimetric analysis. For example, a calcium salt in a solution made acidic with ethanoic acid, when treated with an excess of sodium oxalate solution, precipitates calcium oxalate monohydrate quantitatively [52]. The precipitate is washed with ethanol, and may then be analysed as one of three compounds:

- as $CaC_2O_4 \cdot H_2O$, by drying at 100–115 °C for 1–2 hours;
- as $CaCO_3$ by heating to 475–525 °C in a furnace;
- as CaO by igniting at 1200 °C.

When we heat this compound on a thermobalance, we obtain the curves shown in Figure 2.15. A single run in air at 10 °C/min up to 1000 °C shows three distinct changes:

1. Between room temperature and about 250 °C there is a loss of about 12% to give a stable product.
2. Between 300 and 500 °C a loss of some 19% occurs to give a second stable product.
3. Above 600 °C there is a final loss of about 30% to give a final product stable to the highest temperature used.

Can we suggest a set of reactions to explain these losses?

1. Since the first loss starts around 100 °C, we might consider loss of water vapour:

$$CaC_2O_4 \cdot H_2O(s) \quad = \quad CaC_2O_4(s) + H_2O(v)$$
Calculated % loss $= 100 \times 18.0/146.1 = 12.3\%$

This loss corresponds with what is found, so this reaction is *possible*.

2. Previous analytical methods produced calcium carbonate by heating to about 500 °C, so let's consider that:

$$CaC_2O_4(s) \quad = \quad CaCO_3(s) + CO(g)$$
Calculated % loss $= 100 \times 28.0/146.1 = 19.2\%$

This, too seems a likely reaction!

3. The decomposition of calcium carbonate is a well-established reaction:

$$CaCO_3(s) \quad = \quad CaO(s) + CO_2(g)$$
Calculated % loss $= 100 \times 44.0/146.1 = 30.1\%$

Words of caution! Despite the excellent agreement of the calculated and experimental values, it is *most unwise* to deduce reaction schemes from this evidence alone! The products, both gaseous and solid, should be characterised by other analytical methods, such as X-ray diffraction for solids and chemical tests for gases.

In the particular case of calcium oxalate described above, these tests have been made, and the stages of decomposition confirmed. The presence of calcium oxalate hydrate in kidney stones [53] extends the use of this analysis to biological fields.

If the conditions of analysis are changed, the temperature ranges of the events may alter, and the shape of the curve change, but the stable products will still be obtained. We shall see later that this analysis may be adapted to determine calcium, strontium and barium simultaneously [52].

COPPER SULPHATE PENTAHYDRATE $(CuSO_4 \cdot 5H_2O)$

Blue hydrated copper sulphate is a very well known material and has been used to demonstrate many thermal techniques. In this example, we shall

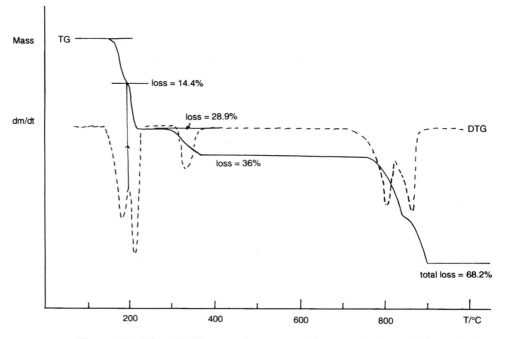

Figure 2.16 TG and DTG curves for copper sulphate pentahydrate, 10.61 mg, alumina crucible, 10 K/min, air.

show how the DTG curve may be used to help obtain quantitative data. Depending on the conditions, the stages of the reactions may be resolved to different extents. If a very fast heating rate and large sample is used, the stages tend to 'blur' together. Conversely, at slow heating rates, or where the rate of reaction is controlled, the stages are well resolved.

The sample heated for Figure 2.16 shows a mass loss starting below 100 °C, closely followed by a further loss up to 160 °C. The DTG curve shows a minimum from which the mass losses for the two overlapping stages may be calculated, using the construction shown.

By checking the mass losses, you can confirm the following *possible* set of decomposition reactions. (RAM are: Cu = 63.55; S = 32.06; O = 16.00; H = 1.00.)

1. $CuSO_4 \cdot 5H_2O(s) = CuSO_4 \cdot 3H_2O(s) + 2H_2O(v)$
2. $CuSO_4 \cdot 3H_2O(s) = CuSO_4 \cdot H_2O(s) + 2H_2O(v)$
3. $CuSO_4 \cdot H_2O(s) = CuSO_4(s) + H_2O(v)$
4. $CuSO_4(s) = CuO(s) + SO_2(g) + \frac{1}{2}O_2(g)$

What does the DTG curve suggest about the reaction above 700 °C? One explanation given is that a basic sulphate, $CuO \cdot CuSO_4$, is formed.

DEGRADATION OF POLYMERS

The effects of heat on polymeric materials varies greatly with the nature of the polymer, ranging from natural polymers like cellulose, which start to decompose below 100 °C, to polyimides which are stable up to 400 °C. The presence of additives and the atmosphere used also affect the changes greatly.

When heated in an inert atmosphere, polymeric materials react by two general routes, often related to their mode of polymerisation and the heat of polymerisation. They may either *depolymerise* or *carbonise* [4,54].

Polymers such as poly(methyl methacrylate) (PMMA) and polystyrene (PS) may depolymerise or 'unzip' back to the monomer. Such polymers generally have a low ΔH of polymerisation. Random scission of poly-(alkenes) like poly(ethylene) (PE) will produce unsaturated hydrocarbons from chain fragments of varying lengths like butenes, hexenes and dienes. The kinetics of this random process is approximately first order, and the temperatures of degradation are much affected by the substitution. Thus, with fluorine substitution, poly(tetrafluoroethylene) (PTFE) decomposes at a much higher temperature than PE, and with methyl substitution, poly(propylene) (PP) at a lower temperature than PE, as shown in Figure 2.17.

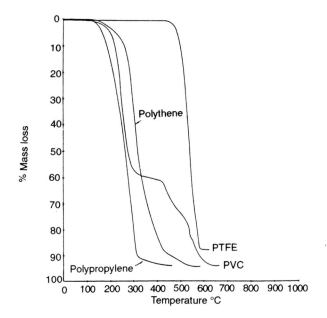

Figure 2.17   TG curves for polymer samples in air. Samples all about 1 mg, heated at 50 K/min.

Poly(vinyl chloride) (PVC) and poly(acrylonitrile) (PAN) may eliminate small molecules initially and form unsaturated links and cross-links before finally degrading by complex reactions to a char, which will oxidise in air.

$$R—(CH_2—CH \cdot Cl—)_n—R' \rightarrow R—(CH=CH)—R' + HCl \rightarrow etc.$$

This stage is shown in Figure 2.17 as a loss of about 60%.

Polymers containing polar groups, such as polyamides (e.g. nylon 66), may absorb moisture which is usually lost below 100 °C, sometimes in stages if the moisture is in different environments.

Cellulose polymers, polyester resins and phenol–formaldehyde polymers have extremely complex decomposition schemes, eliminating small molecules, often flammable or toxic, and eventually leaving a charred mass. If a polymer melts during degradation, a *coke* is formed.

In an oxidising atmosphere, these reactions are further complicated, both by the oxidative attack on the polymer, forming peroxides, and also by the oxidation of the products and char.

Flame-retardant additives in polymers may be studied since their mode of action may alter the nature of the reactions and the amounts of products.

In Figure 2.17, several polymers are compared and their relative stabilities may be assessed. The initial temperature of degradation may not be comparable to the maximum continuous use temperature, since thermoplastics may melt, and this would not show on TG, and other polymers degrade slowly over a long period, even at low temperatures.

A cautionary note on the use of thermal methods of stability assessment was sounded by Still [56], whose review considers instrumental and procedural problems and the effects of polymer structures. Cullis and Hirschler [57] consider thermal methods as applied to the combustion of polymers, and Turi [58] provides a comprehensive review of polymer applications of thermal methods.

2.6.2   *Analysis of mixtures*

When we wish to analyse an unknown mixture of chemical substances in a real sample, several options are open to us. We may use the wide range of separation techniques such as chromatography to isolate each component of the mixture, which may then be identified and measured separately, or we may use a solution technique such as inductively coupled plasma spectrometry which can determine all the elements present simultaneously. However, neither of the above methods is suitable to the entire range of solids from minerals to polymers, and both require a change in the original material before analysis.

Thermogravimetry shows the differences between the behaviours of substances on heating, and if those behaviours are sufficiently different on the temperature scale, the individual reactions of substances may be identified and measured. Three examples will be given from a wide range.

MIXTURES OF ALKALINE EARTH OXALATES

We have already seen how the decomposition of calcium oxalate hydrate may be followed by TG. The oxalates of magnesium, strontium and barium are also precipitated under similar conditions, but decompose in rather different ways [52].

Magnesium oxalate dihydrate breaks down in only *two* stages: firstly, to the anhydrous oxalate by about 250 °C, and then directly to the oxide by 500 °C.

Strontium oxalate monohydrate and barium oxalate hemihydrate lose water by 250 °C and the anhydrous oxalates decompose to the carbonates by 500 °C. The strontium carbonate decomposes between about 850 and 1100 °C while the barium carbonate is stable until even higher temperatures. An illustration of the decomposition of the mixture of calcium, strontium and barium oxalate hydrates is shown in Figure 2.18 and the percentages of each may be found from the TG trace.

A comparison of the thermodynamics of decomposition of metal oxides or metal carbonates (Figure 2.19) may be done by the construction of Ellingham diagrams [59]. These involve the plotting of $\Delta G^{\ominus}$ against T for the reaction of importance, for example:

$$M_mCO_3(s) = M_mO(s) + CO_2(g)$$

For this reaction ΔG and ΔH are generally positive at 298 K (25 °C) and ΔS is very positive since gaseous carbon dioxide is produced. For calcium carbonate, for example, at 298 K:

$$\Delta G^{\ominus} = 130.1 \text{ kJ/mol};$$
$$\Delta H^{\ominus} = 177.9 \text{ kJ/mol; and } \Delta S^{\ominus} = 160.4 \text{ J/(K mol)}$$

The slope of the Ellingham plot is:

$$(\partial \Delta G^{\ominus}/\partial T)_p = -\Delta S^{\ominus}$$

and is negative since $\Delta S^{\ominus}$ is very positive. This means that $\Delta G^{\ominus}$ becomes less positive and finally passes through zero to become negative as the temperature increases. When $\Delta G^{\ominus}$ is zero, the equilibrium pressure of carbon dioxide is 1 atmosphere and the carbonate will decompose. We may call this a 'decomposition temperature T_d', although the material may decompose at other temperatures under other conditions.

It has been shown that the value of T_d depends upon the cation radius, r, and a good correlation is found plotting $\Delta H^{\ominus}$ vs $(r^{1/2}/z^*)$ where z^* is the effective nuclear charge.

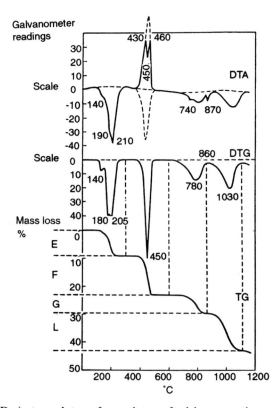

Figure 2.18 Derivatograph trace for a mixture of calcium, strontium and barium oxalate hydrates, about 0.3 g, heated at 10 K/min [52]. The stages are as follows: loss of water of crystallisation (E); loss of CO from decomposition of the oxalates (F) and loss of CO_2 from calcium carbonate (G) and from strontium carbonate (L).

Mixtures of carbonates and mixtures of nitrates may be analysed quantitatively by TG and examples are given at the end of this chapter and in Chapter 6. Dollimore *et al.* [60] have shown the applicability of the thermodynamic treatment to more complex systems.

POLYMER BLENDS

Since the decomposition profiles of polymers are characteristic, some polymer blends [61,62,63] may be analysed by TG. Figure 2.20 shows the TG of a ethylene–vinyl acetate (EVA) copolymer heated at 20 K/min in nitrogen. The first loss may be shown to be ethanoic acid (acetic acid) from the vinyl acetate, while the second shows the decomposition of the residual material.

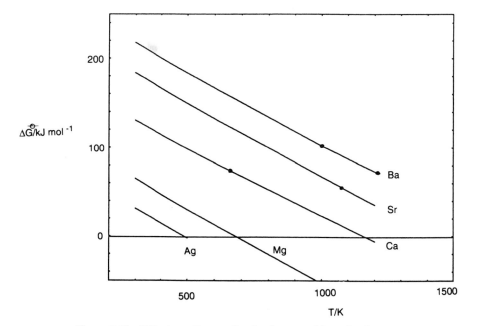

Figure 2.19 Ellingham diagram for the decomposition of carbonates: $M_mCO_3 = M_mO + CO_2$. Full points represent phase transitions [35].

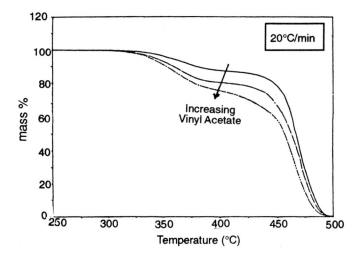

Figure 2.20   TG curves for ethylene–vinyl acetate blends. Samples of about 10 mg run in nitrogen at 20 K/min. (Courtesy TA Instruments Ltd.)

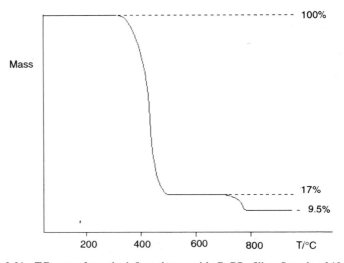

Figure 2.21   TG curve for polyolefin polymer with CaCO₃ filler. Sample of 10.5 mg heated at 10 K/min in nitrogen. Residue at 500 °C gives the filler content (17%) and the temperature and percentage mass loss for decomposition (about 700 °C and 44% loss) suggest CaCO₃.

Polymers containing additives, especially fillers like calcium carbonate and fire-retardants such as aluminium hydroxide, are often studied by TG. The decomposition stages may be altered, but the residue at high temperature is characteristic of the filler. For example, calcium carbonate shows the normal decomposition to calcium oxide around 800 °C and aluminium hydroxide leaves a stable residue of Al_2O_3. Figure 2.21 shows one example of this.

The quantitative analysis of filled polymers and rubbers may also involve oxidation. Figure 2.22 shows a rubber material filled with carbon. The stages of decomposition are well defined, and start in nitrogen atmosphere with the loss of a volatile oil component below 200 °C followed by the degradation of the main polymer components. The residue at 600 °C is stable in nitrogen, but on switching to air the carbon-black filler is oxidised away, leaving the small ash content. We thus have four analyses performed in one TG run. It should be noted that the decomposition products of the polymer are often very flammable and care must be taken to remove them before introducing any air!

SOILS

The composition of soils [2,64,65] is complex, and varies with the geological and biological nature of the area, the level from which the soil

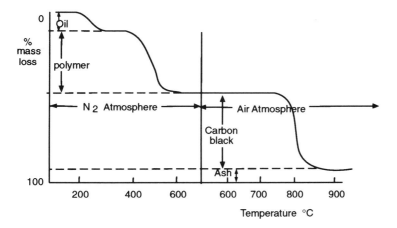

Figure 2.22   TG curve for carbon-black filled rubber. (Courtesy PL-Thermal Sciences.)

was collected and the human and biological interaction that has taken place. After soil samples have been collected, they should be stored under reproducible conditions, e.g. constant relative humidity for about 1 week. There are generally three stages of soil decomposition on heating:

1. Loss of hygroscopic moisture and of simple volatiles produced from organic compounds between room temperature and 150 °C.
2. Ignition of organic matter starts at about 250 °C and is complete by 550 °C. The organic matter may be determined by the mass loss between these two temperatures, provided no other reactions interfere. The largest interference comes from clay minerals, but provided the clay content is less than about 40%, the loss to 500 °C is a very good guide to the organic matter. An alternative is to remove the organics by wet oxidation using hydrogen peroxide, and then run the residue.
3. Hydrated minerals such as aluminium and iron oxides, and micas and gypsum may complicate matters, but high-temperature decompositions can show the content of minerals such as carbonates.

COALS

The analysis of coals [66,67] for moisture, total volatiles, fixed carbon and ash content is referred to as a 'proximate analysis'. Thermogravimetry, especially when computer controlled, has great advantages over traditional methods for proximate analysis. Figure 2.23 shows the TG curve of a typical coal, and the alterations in heating rate and in atmosphere required

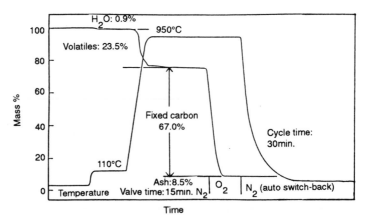

Figure 2.23   Proximate analysis of coal using microcomputer-controlled thermogravimetry [67].

to obtain results in the optimum time. It is most important that each stage is completed before starting the next, especially when switching to an oxidising atmosphere, where any remaining flammable gases may combust explosively at high temperature in oxygen or air.

2.6.3   Oxidation studies

Besides the oxidation of organics and of carbon, studies have been made of the oxidation of metals and alloys [68–70] and of compounds [71] using thermogravimetry.

   In this instance it is often useful to run the sample *isothermally*. Figure 2.24 shows the oxidation of metallic copper in air at 800 °C. The curve is unusual in showing a weight *gain*! Where this is observed in other traces it often indicates metal oxidation. The oxidation of copper forms Cu_2O first. The shape of the curve may be fitted to the parabolic law governing the kinetics of metal oxidation and may show a broken parabola if the oxide surface cracks [72].

2.6.4   Reduction studies

The preparation of catalysts often involves a reduction step. The use of TG apparatus and of special *temperature programmed reduction* (TPR) apparatus, which measures continuously the consumption of hydrogen during the reduction process, has given much useful data on catalyst preparation and allowed the detection of overlapping reactions and the study of kinetics [73,74].

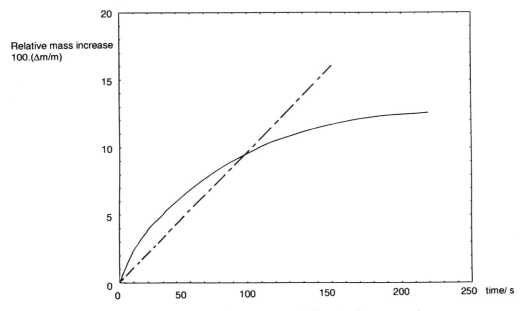

Figure 2.24 Isothermal TG oxidation curve for copper metal. Sample of copper turnings, 20 mg, heated at 800 °C in air. Relative mass increase (full line) and $(\Delta m/m)^2$ (dashed line) for parabolic rate law.

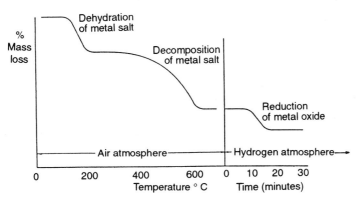
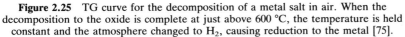

Figure 2.25 TG curve for the decomposition of a metal salt in air. When the decomposition to the oxide is complete at just above 600 °C, the temperature is held constant and the atmosphere changed to H_2, causing reduction to the metal [75].

In a similar way to that described in section 5 above, the atmosphere may be changed during the run so that the product is reduced. This is shown in Figure 2.25.

In the usual TG experiments, the analyst sets a *constant heating rate* and measures the temperature and mass as dependent variables, as a function of time. If several reactions occur, it is important that we try to resolve them. Of course, the atmosphere gases, the instrumental conditions and the nature of the sample, plus the kinetics of the reactions occurring, will affect the extent to which the reactions may overlap. Very small samples will help, producing less interference between products and reactants, and very low heating rates will help by allowing one reaction to finish before the next begins. However, these conditions cause problems, either of sensitivity, or of a very long time needed for the experiment, perhaps several hours if isothermal or at a very slow heating rate.

Another method of improving the resolution is to control the *rate of reaction*. This may be done by sensing the rate of mass loss [48] or by sensing the evolution of a particular product gas [47] and using one of these parameters to control the rate of heating. Generally these methods have led to longer experimental times than for a conventional TG run.

Commercial equipment is now available [76] which achieves the benefit of higher resolution of thermal events without sacrificially long experimental time. With microprocessor control, when no mass loss is occurring, it is possible to have a high heating rate (for example, 50 K/min), but when a loss is detected the heating rate is reduced towards zero, and kept low until the mass again becomes nearly constant. One analogy is with a video recorder. If nothing of interest is happening, we can 'fast forward' the videotape. When an important event shows up, we can slow down or even 'pause' the tape.

For rapid sensing and control, it is necessary to have a thermocouple close to the sample, plus a directed purge gas flow for rapid removal of product gases and interchange with the sample, plus a thermobalance system of high sensitivity to detect small mass changes and respond rapidly. The commercial Hi-Res™ TGA system allows the operator to programme in the resolution required, from a very high resolution where the experimental time may be long, to a 'standard' experiment with a run giving much higher resolution than a normal TG, but a comparable run time.

Applications of Hi-Res™ TGA have been reported for various materials, from improved resolution for the dehydration and decomposition of copper sulphate, to the analysis of fuel transport additives. Three examples showing the improvements that may be obtained will be given.

2.7.1   *Polymer blends*

We have noted above the possibility of analysing ethylene–vinyl acetate (EVA) copolymers by TG (see Figure 2.20). Although the steps are fairly well resolved with conventional TG, Hi-Res™ TGA improves this dramatically and gives better quantitative accuracy.

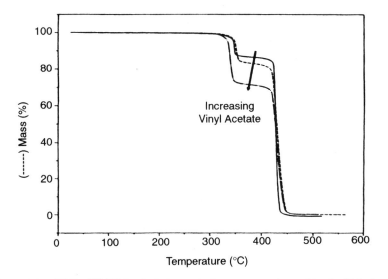

Figure 2.26 Hi-Res™ TGA of ethylene–vinyl acetate copolymer blends. This figure should be compared with Figure 2.20. (Courtesy TA Instruments Ltd.)

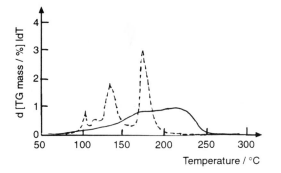

Figure 2.27 Comparison of Hi-Res™ and conventional TG first derivatives for a polymer derivative fuel additive. Solid line, conventional TG; dashed line, Hi-Res™/TG +4.

2.7.2 Fuel additives

Fuel and oil additives [77] may be subjected to an oxidative environment and temperatures around 200 °C. They are designed to improve the properties of the oil blend, especially the flow and oxidative stability. Studies of their thermal stability are most important.

Conventional TG of a polymeric derivative used as an additive shows a gradual, poorly differentiated mass loss. Figure 2.27 shows the TG first derivative curves for conventional and Hi-Res™ TGA of this material. The Hi-Res™ TGA gives a much clearer resolution of the mass losses into three steps, after loss of low-boiling solvent. The mass losses can be quantified more accurately, and the product gases identified by mass spectrometry.

2.7.3 *Drugs*

The presence of water in both 'free' and 'bound' states in pharmaceutical materials is most important. Estimation of the amounts of each type, and the study of water bound in different ways may give better information about the compatibility and action of the drug [78].

Figure 2.28 shows the conventional and Hi-Res™ TGA of a hydrated drug substance, and the DTG curves enable a better separation of the

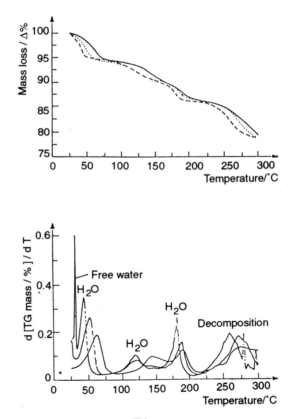

Figure 2.28   Comparison of the Hi-Res™ and conventional TG and DTG of a hydrated drug candidate [78]. Solid line, conventional TG; dotted line, Hi-Res™ index 4; dashed line, Hi-Res™ index 5.

water present in various environments within the material, from the 'free' moisture to two or more kinds of 'bound' water. The trace also shows decomposition at higher temperatures.

2.8 Problems
(Solutions on p. 273)

1. Which of the following physical changes could NOT be detected by thermogravimetry?
   (a) loss of moisture; (b) sublimation; (c) melting; (d) gas adsorption.
2. Which of the following chemical reactions would NOT be detected by thermogravimetry?
   (a) $CaCO_3 + SO_2 = CaSO_4 + CO_2$
   (b) $CaCO_3 + SiO_2 = CaSiO_3 + CO_2$
   (c) $CaCO_3 + Na_2SO_4 = CaSO_4 + Na_2CO_3$
   Calculate the mass change in *any* of the reactions that could be detected.
3. You have decided to investigate a material using thermogravimetry. Make a check-list of all the things you *should* do before attempting to run the sample. If you are investigating the *stability* of the material, what would you do as additional experiments?
4. Magnesium carbonate was dissolved in aqueous oxalic acid and a crystalline product A obtained. When A was heated in air, 9.20 mg of A lost mass in *two* stages only: 2.23 mg were lost up to 220 °C and a further 4.49 mg by 500 °C. Write balanced chemical equations for the preparation and decompositions of A.
5. An alloy containing silver and copper was dissolved in concentrated nitric acid in a small ceramic TG crucible. When heated in air it gave the TG curve of Figure 2.29. Observations showed the residue at the first

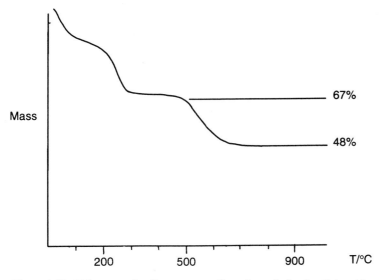

Figure 2.29 TG curve of a silver–copper alloy after solution in nitric acid.

plateau was blue crystals, at the second a black solid and at the third a mixture of black solid and metallic specks. Calculate the percentage silver in the alloy.

6. Using the curves given in Chapter 2 as a guide, sketch the curves for polyethylene (PE) and for poly(vinyl acetate) (PVA). If a copolymer of these two (EVA) loses 15% of its weight in the first stage heating up to 390 °C, calculate the percentage composition of the copolymer.

7. A colleague has performed a series of TG experiments on a sample of a 'pure' substance. He approaches you in some concern, because the results are very different in appearance and in temperatures of the steps. What might he be doing incorrectly? If his results are *correct*, what does this suggest about the sample?

References

1. A.I. Vogel, *Quantitative Inorganic Analysis* (3rd edn), Longman, London, 1961, p. 104.
2. I. Barshad in *Methods of Soil Analysis*, Pt 1, C.A. Black (ed.), No. 9 in Agronomy Series, American Soc. Agronomy, Madison, 1965, Ch. 50.
3. C. Duval, *Inorganic Thermogravimetric Analysis* (2nd edn), Elsevier, Amsterdam, 1963.
4. C.J. Keattch, D. Dollimore, *Introduction to Thermogravimetry*, Heyden, London, 1975, Ch. 1.
5. W. Wendlandt, *J. Chem. Educ.*, 1972, **49**, A571, A623
6. K. Honda, *Sci. Rep. Tohuku Univ.*, 1915, **4**, 97
7. M.I. Pope, M.D. Judd, *Educ. in Chem.*, 1971, **8**, 89
8. P. Chevenard, X. Wache, R. de la Tullaye, *Bull. Soc. Chim. Fr.*, 1944, **10**, 41
9. L. Cahn, H.R. Schultz, *Vacuum Microbalance Techniques*, Vol. 2, Plenum, New York, 1962, p. 7.
10. J.O. Hill, *For Better Thermal Analysis and Calorimetry*, III, ICTAC, 1991.
11. P.D. Garn, O. Menis, H.-G. Wiedemann, *ICTA Magnetic Reference Materials Certificate*, NBS GM-76, 1976.
12. J.M. Thomas, B.R. Williams, *Quart. Rev.*, 1965, **XIX**, 239
13. F.W. Sears, M.W. Zemansky, H.D. Young, *College Physics* (5th edn), Addison Wesley, Reading, MA, 1980.
14. W.L. de Keyser, *Nature*, 1962, **194**, 959.
15. W.-D. Emmerich, E. Kaisersberger, *J. Thermal Anal.*, 1988, **34**, 543; Netzsch TG 439 Brochure, Netzsch Mastermix Ltd.
16. T. Gast, E. Hoinkis, U. Muller, E. Robens, *Thermochim. Acta*, 1988, **134**, 395.
17. Netzsch TG 449 Brochure, Netzsch Mastermix Ltd.
18. E.L. Charsley, A.C.F. Kamp, J.P. Redfern, *Progress in Vacuum Microbalance Techniques*, Vol. 2, eds S.C. Bevan *et al.*, Heyden, London, 1973, p. 97
19. D.A. Skoog, D.M. West, *Fundamentals of Analytical Chemistry* (4th edn), Holt-Saunders, Philadelphia, 1982, Ch. 24.
20. F.W. Fifield, D. Kealey, *Analytical Chemistry* (3rd edn), Blackie, Glasgow, 1990, Ch. 12.
21. F. Brailsford, in *Permanent Magnets and Magnetism*, D. Hadfield (ed.), Iliffe/Wiley, London, 1962.
22. P.K. Gallagher *et al.*, *J. Thermal Anal.*, 1993, **40**, 1423
23. A.R. McGhie, J. Chiu, P.G. Fair, R.L. Blaine, *Thermochim. Acta*, 1983, **67**, 241
24. E.L. Simons, A.E. Newkirk, *Talanta*, 1964, **11**, 549
25. A.E. Newkirk, *Thermochim. Acta*, 1971, **2**, 1
26. H.R. Oswald, H.G. Wiedemann, *J. Thermal Anal.*, 1977, **12**, 147
27. P.D. Garn, *Anal. Chem.*, 1961, **33**, 1247
28. M.L. McGlashan, *Physico-Chemical Quantities and Units*, RSC, London, 1968, p. 39.
29. J. Sestak, G. Berggren, *Thermochim. Acta*, 1971, **3**, 1

30. M.E. Brown, D. Dollimore, A.K. Galwey, *Comprehensive Chemical Kinetics*, Vol. 22, *Reactions in the Solid State*, Elsevier, Amsterdam, 1980.
31. M. Reading, D. Dollimore, J. Rouquerol, F. Rouquerol, *J. Thermal Anal.*, 1984, **29**, 775.
32. M. Arnold, G.E. Veress, J. Paulik, F. Paulik, *Thermochim Acta*, 1982, **52**, 67
33. E. Urbanovici, E. Segal, *J. Thermal Anal.*, 1993, **40**, 1321
34. A.K. Galwey, *Thermal Analysis, Proc 7th ICTA*, Wiley, Chichester, 1982, p. 38.
35. K.H. Stern, E.L. Weise, NSRDS-NBS 30: Carbonates, 1969.
36. A. Bhattacharya, *J. Thermal Anal.*, 1993, **40**, 141.
37. J.H. Sharp, G.W. Brindley, B.N.N. Achar, *J. Am. Ceram. Soc.*, 1966, **49**, 379.
38. L.K. Avramov, *Thermochim. Acta*, 1985, **87**, 47.
39. M. Reading, *Thermochim. Acta*, 1988, **135**, 37.
40. C.D. Doyle, *J. Appl. Polym. Sci.*, 1962, **6**, 639.
41. A.W. Coats, J.P. Redfern, *Nature*, 1964, **201**, 68.
42. M.E. Brown, *Introduction to Thermal Analysis*, Chapman & Hall, London, 1988.
43. T. Ozawa, *Bull. Chem. Soc. Japan*, 1965, **38**, 1881.
44. J.H. Flynn, L.A. Wall, *Polym. Lett.*, 1966, **4**, 323.
45. Perkin-Elmer Analytical report, 1985, p. 16.
46. Netzsch Kinetics Software Brochure.
47. J. Rouquerol, *Bull. Soc. Chim. Fr.*, 1964, 31.
48. F. Paulik, J. Paulik, *Anal. Chim. Acta*, 1971, **56**, 328.
49. P.R. Hornsby, C.L. Watson, *Polym. Deg. Stab.*, 1990, **30**, 73.
50. K.A. Broadbent, J. Dollimore, D. Dollimore, *Thermochim. Acta*, 1988, **133**, 131.
51. V.R. Choudhary *et al.*, *Thermochim. Acta*, 1992, **194**, 361.
52. L. Erdey, G. Liptay, G. Svehla, F. Paulik, *Talanta*, 1962, **9**, 489.
53. E.C. Roberson, Stanton Redcroft: Technical Information Sheet, No 18.
54. F. Rodriguez, *Principles of Polymer Systems* (2nd edn), McGraw-Hill, Singapore, 1983, Ch. 11.
55. Stanton Redcroft TG750 Brochure.
56. R.H. Still, *Brit. Polym. J.*, 1979, **11**, 101
57. C.F. Cullis, M.M. Hirschler, *Polymer*, 1983, **24**, 834
58. E.A. Turi, *Thermal Characterisation of Polymeric Materials*, Academic Press, New York, 1981.
59. H.T. Ellingham, *J. Soc. Chem. Ind.*, 1944, **63**, 125
60. D. Dollimore, D.L. Griffiths, D. Nicholson, *J. Chem. Soc.*, 1963, 2617
61. E.L. Charsley, S. St J. Warne, S.B. Warrington, *Thermochim. Acta.*, 1987, **114**, 53.
62. ASTM E1131–86: *Compositional Analysis by Thermogravimetry*, ASTM, Philadelphia, 1986
63. J. Chiu, in *Thermoanalysis of Fibers and Fiber-Forming Polymers*, ed. R.F. Schwenker, Interscience, New York, 1966, p. 25.
64. M. Schnitzer, J.R. Wright, I. Hoffman, *Anal. Chem.*, 1959, **31**, 440.
65. B.D. Mitchell, A.C. Birnie, in *Differential Thermal Analysis*, Vol. 1, ed. R.C. Mackenzie, Academic Press, London, 1970, Ch. 24, p. 695.
66. S. St J. Warne, *Proc 7th ICTA, Toronto*, 1982, Wiley Heyden, Chichester, 1982, p. 1161.
67. C.M. Earnest, R.L. Fyans, *Proc 7th ICTA, Toronto*, 1982, Wiley Heyden, Chichester, 1982, p. 1260; and *Perkin-Elmer Thermal Analysis Application Study*, #32.
68. O. Kubaschewski, B.E. Hopkins, *Oxidation of Metals and Alloys*, Butterworth, London, 1962.
69. B.O. Haglund, *Proc. 1st ESTA, Salford*, 1976, Heyden, London, 1976, p. 415.
70. G. Baran, A.R. McGhie, *Proc 7th ICTA, Toronto*, 1982, Wiley Heyden, Chichester, 1982, p. 120.
71. Y. Shigegaki, S.K. Basu, M. Taniguchi, *Thermochim. Acta*, 1988, **133**, 215.
72. J.P. Chilton, *Principles of Metallic Corrosion*, Royal Society of Chemistry Monographs for Teachers, No 4 (2nd edn), RSC, London, 1973.
73. N. Pernicone, F. Traina, *Pure Appl. Chem.*, 1978, **50**, 1169.
74. S.D. Robertson, B.D. McNicol, J.H. de Baas, S.C. Kloet, J.W. Jenkins, *J. Catal.*, 1975, **37**, 424.
75. J.G. Dunn, Stanton Redcroft Technical Information Sheet, No. 103, 1977.
76. TA Instruments Ltd, TGA 2950 System Brochure.

77. T.J. Lever, A. Sutkowski, *J. Thermal Anal.*, 1993, **40**, 257
78. A.F. Barnes, M.J. Hardy, T.J. Lever, *J. Thermal Anal.*, 1993, **40**, 499.

Bibliography

Note The general thermal analysis texts given at the end of Chapter 1 have substantial sections on thermogravimetry but are *not* listed again here.

C. Duval, *Inorganic Thermogravimetric Analysis* (2nd edn), Elsevier, Amsterdam, 1963.
C.J. Keattch, D Dollimore, *Introduction to Thermogravimetry*, Heyden, London, 1975.

Differential thermal analysis and differential scanning calorimetry 3

P.J. Haines and F.W. Wilburn

When an organic chemical is prepared, it is checked against known materials for purity and stability. One of the easiest ways of doing this is to determine its melting point! *This cannot be done by TG, since it involves no mass change*! This also applies to solid–solid phase transitions and to some solid-state reactions. We therefore need a technique that identifies these transitions. Such a technique will give us a simple way of identifying and assessing the quality of materials that may be applied across a large range of substances, from polymers to metals.

The magnitude and direction of the heat change of the system is more difficult to measure accurately and requires the more elaborate apparatus of calorimetry. The techniques to be discussed in this chapter give an interface between very sophisticated precision calorimetry and the qualitative observations of melting points and heat changes.

The early work in studying the interplay of heat and chemical or physical changes was largely qualitative, but the experiments of Lavoisier and Laplace around 1780 allowed the measurement of the quantities of heat energy associated with changes of state and chemical reactions. Although their apparatus required large quantities of sample, was slow to use and of low accuracy, Lavoisier's work has earned him the soubriquet of 'the grandfather of thermochemistry' [1]

During the same period, Fourier worked out mathematical descriptions of transfer processes and he is responsible for many of the fundamental ideas and equations that describe heat transfer. The idea for Ohm's law defining resistance came from Fourier's work, and the compliment is repaid in the use of the 'thermal analogue of Ohm's law' in some theories of thermal analysis [2].

The work of Joule in the nineteenth century led to a greater understanding of the relationship of heat and work, and of the laws governing electrical heating, which now underly much of thermal analysis, both in control of the furnaces and in some DSC apparati. Joule also pioneered the use of twin calorimeter vessels, one containing a reference and the other the substance under investigation [3]. Modern high-precision calorimetry apparatus can be large and complex and the calculations involve many corrections [4]. Nowadays, laboratories generally require the opposite of these: small samples, rapid experiments, simplicity of use and ease of obtaining results!

Le Chatelier is often credited with being the first to use a differential thermal technique when he compared the temperatures shown by two thermometers, one within a sample and one outside it, as they were all heated in an oil bath. He was able to record melting transitions and the behaviour of clay samples. He later worked with thermocouples and characterised clay minerals into categories [5].

Roberts-Austen [6] pointed out that if the difference in temperature

$$T_S - T_R = \Delta T$$

is plotted against temperature or time, then a very sensitive method of detecting changes and transitions is obtained. This identifies the technique of *differential thermal analysis* (DTA)

Various theories were proposed to relate the DTA signals obtained to the heat capacity and enthalpy changes of transitions and reactions occurring during heating, but these were not entirely satisfactory. In 1964 Watson *et al.* [7] announced a *differential scanning calorimeter* (DSC) for quantitative measurements. There are now two main types of DSC and the distinction between the various sensors and apparati is given later. It is recognised that DTA and DSC, properly used, can make accurate measurements of the temperatures of thermal events, can detect the exothermic or endothermic nature of the event, and that DSC can make good measurements of the heat changes that occur.

3.3
Definitions

The two techniques will be considered together, since they are frequently used to study the same phenomena (see [8]).

3.3.1   *Differential thermal analysis (DTA)*

A technique in which the *difference in temperature* between the sample and a reference material is monitored against time or temperature while the temperature of the sample, in a specified atmosphere, is programmed.

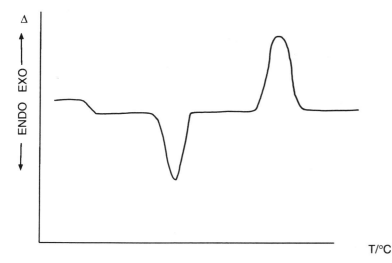

Figure 3.1 Typical DTA or DSC curve, using the convention that endothermic peaks go down. Δ indicates the differential signal for either temperature or power.

The DTA curve is generally a plot of the difference in temperature (ΔT) as the ordinate against the temperature T (or occasionally, time) as the abscissa. An *endothermic* event gives a *downward* 'peak' (Figure 3.1).

3.3.2 *Differential scanning calorimetry (DSC)*

A technique in which the *difference in heat flow* (power) to a sample (pan) and to a reference (pan) is monitored against time or temperature while the temperature of the sample, in a specified atmosphere, is programmed.

In practice, the heat is supplied to the sample contained in the pan, and similarly, to the reference in its pan.

Two types of DSC are recognised:

1. *Power-compensated DSC*, where the sample and reference are heated by separate, individual heaters, and the temperature difference is kept close to zero, while the difference in electrical power needed to maintain equal temperatures ($\Delta P = \mathrm{d}(\Delta Q)/\mathrm{d}t$) is measured.
2. *Heat flux DSC*, where the sample and reference are heated from the same source and the temperature difference ΔT is measured. This signal is converted to a power difference ΔP using the calorimetric sensitivity.

A constant calorimetric sensitivity is desirable, but not essential [9].

The DSC curve has ΔP as the ordinate and temperature (or occasionally, time) as the abscissa. Since an endothermic peak involves the absorption of *more* power by the sample, one convention plots endothermic peaks

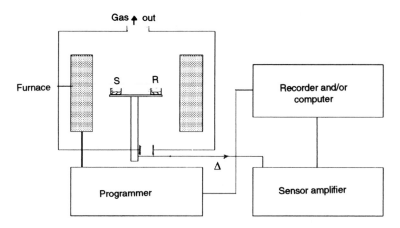

Figure 3.2   Schematic of a DTA or DSC apparatus. Δ indicates the differential signal.

upwards (Figure 3.1). This does cause some confusion, and in this text, we shall adopt the 'DTA' convention throughout, while indicating the sign of heat flow on the *y* axis.

3.4 Apparatus

The major parts of the system (Figure 3.2) are:

1. the DTA or DSC sensors plus amplifier
2. the furnace and its temperature sensor
3. the programmer or computer
4. the recorder, plotter or data acquisition device.

The range of designs for DTA and DSC is extremely wide, and shows a gradation from qualitative DTA to power-compensated DSC, as shown in Figure 3.3 [9].

3.4.1   *The sensors*

As shown in Figure 3.3, thermocouples are used as the sample and reference sensors for many DTA and DSC units. For low temperatures, copper–constantan or chromel–alumel have been used, while for higher temperatures, or more aggressive environments, Pt–Pt/13%Rh has been employed. Single thermocouples are in contact with the sample in ((a), (b)), but outside in types ((c)–(e)). Some Mettler and Setaram DSC units use multiple thermocouples or thermopiles ((d),(e)) to increase the signal. The 'Boersma' DTA, where the heat is conducted to the pans via a conducting metal disc, is used in several apparati.

    The exceptional case is the power-compensated DSC (f), where the

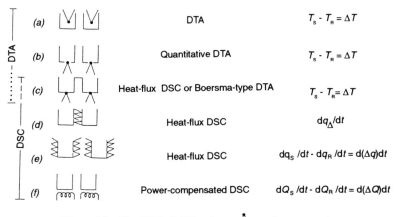

Figure 3.3 The DTA–DSC series. $\bigwedge$, thermocouple; VVV , thermopile; pm , heater [9].

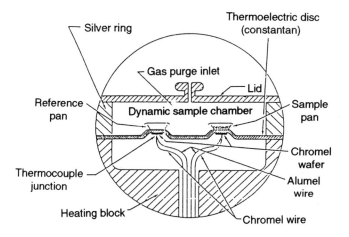

Figure 3.4 Heat flux DSC. (Courtesy TA Instruments Ltd.)

sensors are platinum resistors and the power is supplied to sample and reference separately.

Where the dividing line should be drawn between DTA and DSC is debatable [9], but several different manufacturers have shown [10] that their instruments satisfy the requirements of DSC by determining calorimetric data with good accuracy (Figures 3.4 and 3.5).

Pans and crucibles of many materials have been employed, but the majority of low-temperature instruments use aluminium pans and lids, provided they are not attacked by the samples and are only used well below the melting point of aluminium, 660 °C! For more aggressive environ-

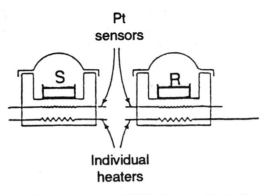

Figure 3.5   Power-compensated DSC. (Courtesy Perkin-Elmer Ltd.)

ments, platinum or ceramic crucibles may be used. The conductivity of the crucibles and their contact with the sensors affect the thermal analysis curves obtained.

A standard experiment might involve a sample of about 10–20 mg of powder, or a disc punched from a polymer film, or a bundle of fibres being placed in a weighed, lidded aluminium crucible, and the total mass recorded. Sometimes an inner liner is placed on top of the sample to enable the sample to make better contact with the base. The lid may be fixed to the base using a simple press, and can give a sealed crucible capable of withstanding about 2 atmospheres pressure. If vaporisation is not a problem, small pinholes can be punched in the lid to allow escape of gas products and reaction with the surrounding purge gas.

Special crucibles have been designed for containing higher pressures [11], or for mixing liquids [12], or for observing the sample during heating [13,14]. Liquids may be injected into the crucible using a syringe.

For the use of even higher pressures, the technique of *pressure DSC* (PDSC) has been invented. The entire DSC cell is enclosed in a strong stainless steel container capable of withstanding pressures from 1 Pa to 7 MPa (approximately 70 atm.). This means that reactions that produce high pressures of gases or react with gases under high pressure may be studied. Additionally, vaporisations may be suppressed by running at higher pressures. Uses of PDSC include the accelerated oxidative stability testing of materials such as oils under high pressures of oxygen [15] and the catalytic reduction of organic compounds with hydrogen [16].

3.4.2   *The furnace and controller*

Very similar comments can be made to those relating to TG (see p.28). For high-temperature DTA, large ceramic-lined, electrically heated furnaces are used with electronic control. Many small DSC systems use a resistance-heated furnace enclosure of silver. The very high thermal conductivity of silver ensures that there is a uniform temperature. For the simplest DTA [17], a metal block, wound with an insulated heating element and having two wells for sample and reference cells and thermocouples, gives a satisfactory introduction to the principles of DTA.

Heating rates between 0 and 100 K/min are used, but the normal rate is about 10 K/min.

One extra feature is the use of DTA or DSC below room temperature. A cooling accessory or refrigeration unit is fitted around the cell and the whole is cooled directly with liquid nitrogen or other coolants. It is essential that dry purge gas is passed through the cell assembly during the cooling, since condensation of water or ice onto the cells might otherwise occur.

3.4.3 *The computer and display*

The need for computer control and rapid data processing is paramount with DSC, since the signal must be converted to the ΔP signal, using a calorimetric sensitivity stored in software, and the curves that are obtained must be analysed for thermal parameters by differentiation (to obtain onset temperatures, for example) and by integration (to obtain peak areas).

3.4.4 *The reference material*

Both DTA and DSC are defined as *differential* methods, where the behaviour of the reference material or pan is compared with that of the sample and its pan. We must take account of ALL the thermal properties that might be involved. For example, the emissivity of the sample may alter if it changes phase, or reacts, or changes colour [18]. This is often avoided by covering both sample and reference pans with snugly fitting lids, or by sealing them. In many cases, small samples may be run on a DSC against an empty pan as reference, but often better results are obtained by using an inert reference material of similar thermal properties in the reference crucible. For many applications, pure, dry pre-heated ('calcined') alumina, Al_2O_3, is satisfactory. Carborundum, SiC, has also been used.

Occasionally, to make the thermal properties of sample and reference more comparable, the sample may be diluted with the reference. It is obviously important that they should not react, and this dilution sometimes improves the baseline and peak shape in DTA experiments.

**3.5
Theory of DTA and
DSC** (F.W. Wilburn)

The shape and size of a typical DTA or DSC curve is determined as much by the environment surrounding the sample and reference materials as by

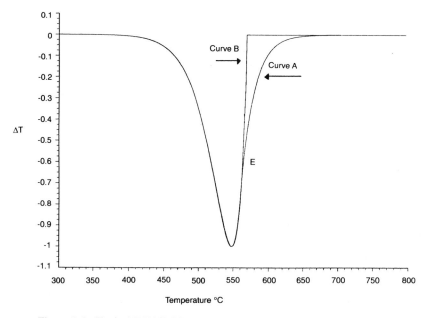

Figure 3.6 Typical DTA/DSC curves: A, practical curve; B, ideal curve.

the mechanism controlling the reaction and the sample material characteristics. Figure 3.6 shows a typical DTA or DSC curve for a material which melts (curve A) together with that (curve B) usually obtained in practical situations. During a melt, the reaction should end at the peak and the curve should then return abruptly to the baseline as in curve A. More often the curve obtained is that of curve B showing a relatively slow return to the baseline.

There are often further complications, in that the signal does not necessarily return to the *original* baseline, but to some other arbitrary baseline above or below this line, as shown in Figure 3.7. Indeed, the situation may be even more complicated, as when the baseline not only has a different value, but also a different slope. All these effects can be explained from the theory of DTA and DSC.

A full discussion of the theory of DTA and DSC would not only require this chapter, but this entire book, and even then it is questionable whether such information would be of use in answering some of the more usual questions asked when gaining information on typical DTA or DSC curves. It is the intention here to explain some of the more common variations seen in experimental curves in terms of theory and, where possible, to suggest ways of reducing errors due to such variations.

We shall first consider the theory of DTA and of the closely related *heat flux DSC*, and extend this to *power-compensated DSC* later.

Theories of DTA can usually be placed into one of two types.

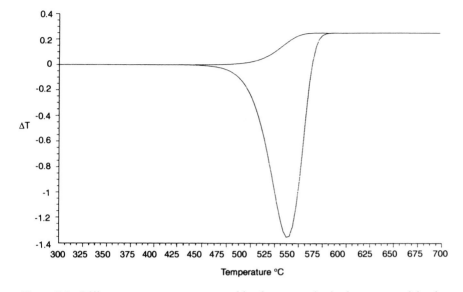

Figure 3.7   Difference temperature measured by thermocouples in the test materials when a change of sample thermal conductivity occurs.

In the first type, heat transfer equations such as the thermal analogue of Ohm's law:

$$dq/dt = (T_a - T_b)/R_{ab}$$

– where dq/dt is the heat flow between a and b and R_{ab} is their 'thermal resistance' – are developed but no account is taken of the reaction mechanism and the part it might play in the production of the peak. Further, for simplicity, it is generally assumed that the different parts of the apparatus (e.g. the sample holder) have no thermal gradients within themselves, which is not necessarily true.

In the second type, reaction equations such as those given by Brown [19] and Keattch and Dollimore [20] are manipulated to develop various relationships between the difference in temperature, ΔT, and the sample temperature at a specific time, from which it is claimed that activation energies and orders of reaction may be obtained. It is questionable whether such relationships are valid when heat transfer is taken into account. Some years ago, an electrical analogue of a DTA was developed [21–24] and was followed with a computer program which allowed heat transfer *and* reaction equations to be combined to show the influence of heat transfer on the shape of DTA curves and its effect on peak area.

Borchardt and Daniels [25] developed heat transfer equations for a system consisting of test tubes of liquid reactants within a stirred, heated liquid. In such a situation the transfer of heat to the test materials would be rapid. Their theory, of the first type, is only valid if a number of limiting conditions are met:

1. The temperature in the holders must be uniform. While this would be true for liquids, it is not the case for solids.
2. Heat transfer must be by conduction only.
3. The heat transfer coefficient must be the same for both holders.
4. The heat capacity of the test materials must be the same.

All these conditions are relatively easy to achieve for liquids, due to the rapid transfer of heat within the system.

Coats and Redfern [26] adopted the second type of approach in that they manipulated reaction equations to show that various relations exist between the peak temperature and the heating rate from which it is possible to derive activation energies and 'orders of reaction'. No account was taken of heat transfer. Nearly all papers adopt one or other of these approaches.

Many earlier pieces of DTA apparatus were made mainly of refractory materials where heat transfer was comparatively slow. As might be expected resulting DTA curves from different apparati differed widely! One of the earliest papers on theory was that of Vold [27] who showed mathematically that there is a relationship between the 'active area' of a DTA curve and the heat involved. The 'active area' is that generated by the reaction itself, while the remaining area is that produced as ΔT returns to the baseline after the reaction has ceased (at point E in Figure 3.6). As this return is governed by natural cooling or heating, which is dependent on the thermal arrangement of the holder system, its form is exponential. Thus, a plot of $\ln(\Delta T)$ vs time for that part of the DTA curve beyond the peak becomes linear after point E. The area between the curve and its baseline from the commencement of the peak to the point E is the 'active area'. Further, Vold showed that the relationship between this 'active area' and the heat of reaction is:

$$\text{Heat} = A \cdot (\text{'active area'})$$

where A is the slope of the linear part of the $\ln(\Delta T)$–time curve. In this theory it was assumed that the physical properties of the sample and reference materials remained constant throughout the reaction. Thus, the shape of a DTA peak, particularly that portion at temperatures above the peak temperature, can be influenced by apparatus parameters. For reactions other than meltings, the assumption that the reaction ends before the DTA curve returns to the baseline is dependent on the type of reaction under investigation. The 'active' part of the DTA curve for melts, crystalline changes and zero-order reactions usually ends around the maximum of the DTA peak, so that the return to the baseline is exponential for a considerable portion of the DTA curve, and thus the end of the reaction can be defined fairly accurately as well as the A factor referred to above. However, if the reaction is controlled by other types of mechanism then ΔT is often close to the baseline before the reaction ends. In this case it is difficult to determine the constant A. The high temperature end of a DTA curve can comprise a mixture of reaction mechanism and exponential return and can thus be different for different apparati.

The fact that this end of the DTA peak may not be a true representation of the reaction can have other consequences if two reactions follow in quick succession. In certain circumstances the two peaks may merge to produce a single peak. Indeed, this was used some years ago to test the discrimination ability of apparatus. A mixture of silica and potassium sulphate has two peaks, one the result of the structural inversion in silica at 573 °C, the other at 583 °C owing to a crystal transition in potassium sulphate. The ability to separate these peaks gave a measure of the resolution of the apparatus.

Many of the theories developed for the examination of DTA traces assumed that both before and after a reaction that produced a peak, the physical properties remained the same and were independent of temperature. On this assumption, the baseline remains the same across the DTA peak. However, there are very few practical DTA (or DSC) curves for which this is so. In older apparatus, the sample and reference temperatures were measured using thermocouples in the centre of the test materials (Figure 3.3(a)), which is the classsical DTA design. With such an arrangement, theory indicates that the position of the baseline is influenced by the relationship between the physical properties of the sample and reference materials. In order to maintain a constant, near-zero baseline, the physical properties such as the thermal conductivity k, heat capacity C and density ρ of sample and reference had to be closely matched. However, if any of these changed, either with reaction or with temperature, then the baseline of the DTA curve varied from zero, as shown in Figures 3.7 and 3.8. It can be shown that, for cylindrical samples of radius r, the equilibrium baseline offset ΔT is:

$$\Delta T = \{\beta \cdot r^2/4\} \cdot [(\rho_R \cdot C_R/k_R) - (\rho_S \cdot C_S/k_S)]$$

where β is the heating rate (K/s).

Usually the greatest change in any property of the sample material occurs during a reaction, so that it is not uncommon for the greatest baseline changes to occur during this period, i.e. during the production of a DTA peak. This change in baseline level affects the calculation of the area of the DTA peak. The magnitude of the error in calculation will depend on the size of the peak, a large peak being less affected than a small one. The error is compounded by not being able to define the 'true' baseline across the peak. The simplest expedient of drawing a straight line across the peak from the point where the DTA curve deviates markedly from the initial curve to a point where the curve returns to a level baseline nearly always results in an area that is too large, as shown in Figure 3.8. The baseline

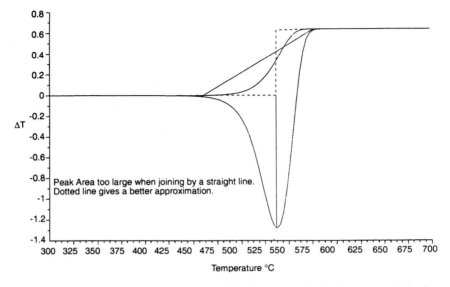

Figure 3.8 Difference temperature measured by thermocouples in the test materials when a change in sample specific heat capacity occurs.

change in Figure 3.8 was computed from preset changes in the physical constants occurring as the reaction was proceeding. It is sometimes possible to define a baseline across the peak in the situation when the level of the baseline is different before and after a peak, even if the true baseline cannot be defined easily [28], and this is shown as the dotted line in Figure 3.8. Much depends on the type of reaction involved and no general rule can be stated.

3.6
Heat flux DSC

If the measurement thermocouples are positioned beneath the sample and reference pans (Figure 3.3(c)–(e)) we have the heat flux DSC designs. The shift in baseline is now only influenced by changes in the sample specific heat capacity (Figures 3.9, 3.10) and is not affected by the other properties of the sample. Such a design of DSC may be used to measure specific heat capacities. Unfortunately, the change in baseline is also dependent on the characteristics of the pans holding the test materials so that the apparatus has to be calibrated using known standard materials.

The area under the peak is directly proportional to the heat of reaction [23,24,26].

$$\Delta H = K \int \Delta T \, dt = K \cdot (\text{peak area})$$

The calibration constant K converts peak area into joules, and is a *thermal factor* which may vary with temperature.

Problems of measurement of the area still arise owing to the difficulty in

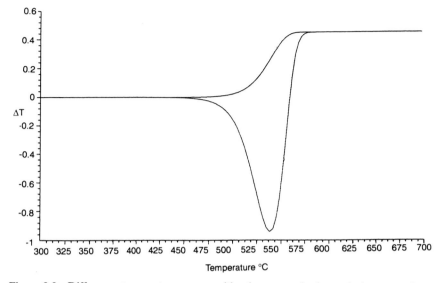

Figure 3.9 Difference temperature measured by thermocouples beneath the pan and test materials when a change in sample specific heat capacity occurs.

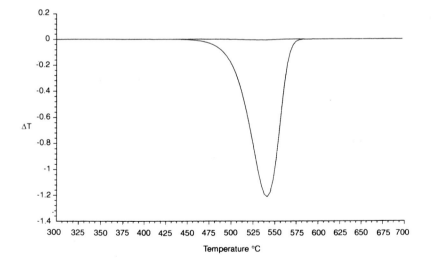

Figure 3.10 Difference temperature measured by thermocouples beneath the pan and test materials when a change in sample thermal conductivity or density occurs.

defining an accurate baseline. However, the DSC design does have some advantages (not least the reduction in chemical attack) over earlier DTA designs where the thermocouples were placed within the sample. Although the sensitivity is greater with the thermocouples within the sample, baseline shifts are often greater, being produced not only by changes in specific heat capacity but also by changes in thermal conductivity and density. In the DSC design, the positioning of the measuring thermocouples outside the test materials may reduce the sensitivity and overall response and some small inversions may not be seen when using certain types of DSC design.

3.7 Power-compensated DSC

In the other type of DSC, the sample and reference holders are insulated from each other and have their own individual sensors and heaters. The electrical circuitry operates to maintain the holders at the same temperature, within the electrical sensitivity of the circuitry, by varying the power supplied to the heater in each of the holders [29,30]. The thermal energy absorbed by the sample per unit time is exactly compensated by the differential electrical power ΔP supplied to the heaters. Measuring this power is equivalent to measuring the thermal power, and thus the baseline is:

$$\Delta P = \mathrm{d}\Delta q/\mathrm{d}t = \beta \cdot (C_S - C_R)$$

and the peak area gives ΔH directly. In power-compensated DSC the calibration constant required to convert peak area to joules is a constant *electrical* conversion factor.

3.7.1 *The effect of higher temperatures*

Most DSC designs developed for the measurement of specific heat capacity and ΔH are most suitable for temperatures up to around 700 °C. Above this, radiation, which is always present but increases with temperature, becomes significant. Most theories of DTA are based on heat transfer by conduction and assume that radiation is insignificant, and so are only relevant at lower temperatures. High-temperature cells tend to follow the DSC design to avoid thermocouple contamination at temperatures up to 1500 °C. However, at high temperatures heat transfer from the source of heat to the sample is rapid and if the measuring thermocouples are located beneath the pans, the sensitivity may be relatively low.

3.7.2 *Sample size*

When a sample is undergoing a reaction, there is, of necessity, a temperature gradient within it, as heat is abstracted (assuming an endo-

Table 3.1 Calibration materials for DTA and DSC [8, 30]

Material		Temperature (°C)	Enthalpy (J/g)
Cyclohexane	(t)	−83	
	(m)	7	
1,2-Dichloroethane	(m)	−32	
Phenyl ether	(m)	30	
Biphenyl	(m)	69.3	120.41
o-Terphenyl	(m)	58	
Polystyrene	(T_g)	105	
Potassium nitrate	(t)	128	
Indium	(m)	156.6	28.71
Tin	(m)	231.9	56.06
Potassium perchlorate	(t)	300	
Zinc	(m)	419.4	111.18
Silver sulphate	(t)	430	
Quartz	(t)	573	
Potassium sulphate	(t)	583	
Potassium chromate	(t)	665	
Barium carbonate	(t)	810	
Strontium carbonate	(t)	925	

Note. (t) = crystal transition; (m) = melting; T_g = glass transition temperature.

thermic change). To obtain meaningful calorimetric data, it is imperative that the sample size be kept to a minimum in order to reduce such gradients.   However, if the sample consists of a number of materials, as in a study of reaction processes, then it is usually difficult to obtain a small, representative sample. In such a situation, the ideal solution is to use a larger sample with the thermocouple embedded in it and to recognise that the apparatus is then only *qualitative* when used in this way.

3.8 Calibration

For accurate work, it is essential to calibrate the temperature scale, and for DSC instruments the apparatus must also be calibrated for calorimetric sensitivity. ICTAC have approved a set of standard substances, which are listed in Table 3.11 together with some other calibrants.

   The particular instrument may have calibration factors already included in the computer software, but these must be checked from time to time. For a pure metal, such as 99.999% pure indium, the melting of which is shown in Figure 3.11, the extrapolated onset temperature T_e should correspond to the correct melting point of the metal, 156.6 °C. The integrated area of the peak A_S may then be used to calculate the calorimetric sensitivity constant K:

$$K = \Delta H_S \cdot m_S / A_S$$

where ΔH_S is the enthalpy of fusion of indium (28.71 J/g), m_S is the mass of sample of indium (g), A_S is the peak area (cm^2) and K is the calorimetric sensitivity (J/cm^2).

Note: If the area is calculated in K·s or W·s then obviously the units of *K* will change!

Calibration with several materials over the entire range of operation should enable the temperature dependence of *K* to be found.

Example   *A sample of 6.68 mg of high purity indium gave a peak of area 21.94 cm². Calculate the value of K at 156 °C.*

$$K = (28.70 \times 6.68 \times 10^{-3})/\, 21.94 = 8.74 \times 10^{-3} \text{ J/cm}^2$$

Comparison of the calorimetric calibration for one instrument, using a wide variety of materials over the range 0–660 °C [31], has shown that there is good agreement, provided samples are of reasonable size, and the heating rate is the same for calibration and measurement. The ICTAC committees on standardisation are continuously checking and reviewing the data and materials available for calibration [32].

A REMINDER!

The curves obtained for DTA and DSC will depend on the samples and instrument conditions used:

Sample:	chemical nature, purity, history.
Crucible:	material, shape.
Rate of heating	
Atmosphere:	gas, static, flowing.
Mass of sample:	volume, packing, distribution, dilution.

**3.9
Applications**   There are such a large number and variety of applications of DTA and DSC that readers should consult the journals or specialist texts for particular examples. The selection given below is representative of the range of uses of the techniques.

The applications may be divided roughly into two categories:

- *Physical changes and measurements*, such as melting, crystalline phase changes, changes in liquid and liquid crystalline states and in polymers, phase diagrams, heat capacity, and glass transitions, thermal conductivity and diffusivity and emissivity.
- *Chemical reactions* such as dehydrations, decompositions, polymer curing, glass formation and oxidative attack.

3.9.1   *Physical changes and measurements*

MELTING POINT AND ΔH_{fusion} OF MATERIALS

The determination of the melting point may be very easily done with simple apparatus, but ΔH of fusion is much more difficult to measure.

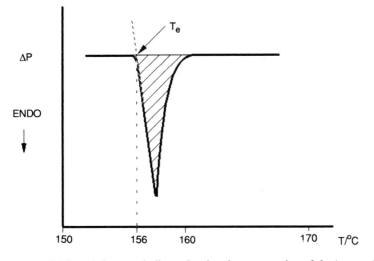

Figure 3.11   DSC peak for pure indium, showing the construction of the 'extrapolated onset' (6 mg, 10 K/min, nitrogen).

While DTA will give excellent qualitative measurements of T_m, we must use DSC for ΔH. Sharp melting peaks, similar to Figure 3.11, have been obtained, with suitable apparatus, over a very wide range of temperature, e.g. from n-heptane at $-90\,°C$ to palladium at $1550\,°C$. The melting points and ΔH of fusion of these compounds determined by DSC or DTA as the onset temperature are in good agreement with the literature values.

The crystallisation of pure materials can be awkward, since they often supercool past their true freezing point by many degrees.

CRYSTALLINE PHASE TRANSITIONS

The crystalline form of a compound can greatly affect its properties, such as solubility, density and electrical properties.

If a substance possesses two or more crystalline forms it is said to be *polymorphic*. If the forms are stable over particular temperature ranges, and have definite transition temperatures, the system is *enantiotropic*. If one form is stable but the other metastable over the whole range of temperature, then the system is *monotropic* and the less stable form will always be tending to transform to the more stable.

POTASSIUM NITRATE, KNO_3

Potassium nitrate shows rather complex behaviour when heated and cooled [31,32]. On first heating, there is an endothermic crystal transition

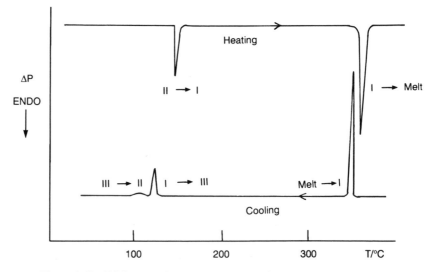

Figure 3.12   DSC curves for potassium nitrate (4 mg, 5 K/min, nitrogen).

(from form II to form I) at 128 °C with a ΔH of about 5.0 kJ/mol and a melting of form I at 334 °C with $\Delta H \sim 10$ kJ/mol. However, if the sample is then cooled from about 150 °C, we obtain an *exothermic* peak at 120 °C with ΔH of only −2.5 kJ/mol. This must mean that it is forming a *different phase. form III.* At lower temperatures it transforms slowly back to form II. These transformations are shown in Figure 3.12.

Other nitrates also show crystal phase transitions, especially rubidium nitrate [33], which has three transitions plus a melting, and ammonium nitrate which has four before melting. Since NH_4NO_3 is widely used as a fertiliser, and occasionally as an explosive, these have been extensively studied [34,35]. Figure 3.13 shows the DTA curve of ammonium nitrate.

First there is a small endotherm around 40 °C from form IV (rhombic II) to form III (rhombic I), followed by a second small endotherm at 80 °C transforming to form II (tetragonal). Two further, larger, endotherms follow at about 120 °C to form I (cubic) and at 170 °C to the melt. These transitions are very dependent on the treatment of the sample, and care must be exercised, since the *very* large exotherm following at about 200 °C represents explosion of the NH_4NO_3!

POLYMORPHISM IN FOODS AND PHARMACEUTICALS

Polymorphism is of particular importance in foods and pharmaceutical preparations, since it can affect the solubility, stability, physiological activity and bioavailability of the compound. As an example we shall

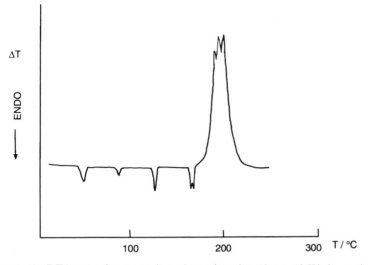

Figure 3.13   DTA curve for ammonium nitrate (powder, 10 mg, 10 K/min, static air).

consider the glycerides, esters of glycerol ($CH_2OH \cdot CHOH \cdot CH_2OH$) and long-chain aliphatic or unsaturated acids. They are present in fats, margarine and chocolate, and are used in many pharmaceutical preparations.

The polymorphism of glycerides has been reviewed by Chapman [36], who notes the use of microscopy, dilatometry, dielectric and X-ray diffraction studies in addition to thermal work and spectroscopy, and Aronhime [37] who concentrates on the applications of DSC. The glyceride systems show at least three phases, distinguished by their spectra and X-ray spacings. The highest melting β form is the most stable. The lowest melting α form and the intermediate β' form are metastable with respect to the β form. Typical DSC traces are shown in Figure 3.14 for tristearin (glyceryl tristearate). On first heating of a sample rapidly cooled from the melt, we observe an initial melting of the α form at around 54 °C almost immediately followed by an exotherm due to the transformation and crystallisation of the stable β polymorph. The β form then melts at about 73 °C. If the liquid is then cooled to a temperature a few degrees above the melting point of the α polymorph, and held there until crystallisation occurs, the β' polymorph is obtained. Heating this form produces a similar trace for the partial melting of β', transformation to β and melting of β, as shown in the dashed curve of Figure 3.14. DSC studies have been made of the polymorphic changes in mixtures of confectionery fats [38] and of the thermal characterisation of edible oils and fats [39].

Polymorphism of pharmaceuticals is discussed fully in the book by Ford and Timmins [40] and by Hardy [41]. The forms of norfloxacin, a derivative of 4-quinolone carboxylic acid, were studied by DSC, TG, X-ray diffraction

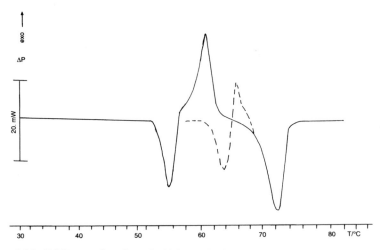

Figure 3.14 DSC curve for tristearin (4.3 mg, 5 K/min, nitrogen). Full line: first heat; dashed line: reheat from 57 °C.

and spectroscopy [42], showing the need to use *all* available analytical methods fully to characterise a material. Two different forms of this drug, A and B, which show a wide spectrum of antimicrobial activity, were obtained from different sources. The forms showed different X-ray patterns and gave the DSC traces in Figure 3.15 when heated in dry nitrogen at 10 K/min. The form A had a melting point of 219.5 ± 0.2 °C with a ΔH of 115 J/g, whereas form B also showed a broad transition at 195.6 ± 0.2 °C with a ΔH of 20 J/g which varied with heating rate. Since this transition was not seen on re-running the sample after cooling, the system is monotropic.

LIQUID CRYSTALLINE TRANSITIONS

The liquid crystalline state is of great importance in the preparation of display devices. Many types of liquid crystal phases are formed from molecules, often with polar end-groups, and with a rod-like structure. The liquid crystals represent various degrees of 'order' within the liquid. From a very ordered solid, we may progress to one of various *smectic* phases, which have plate-like order, or to a *nematic* phase, which has rod-like order, or to the *cholesteric* phase, which is a twisted nematic, and finally to the *isotropic*, completely disordered liquid [43,44].

Considering the order changes in these transitions, the greater the increase in *disorder*, the greater the entropy change ΔS. Since

$$\Delta S = \Delta H/T$$

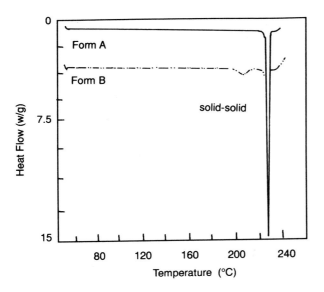

Figure 3.15   DSC curves for two polymorphic forms of norfloxacin. (Redrawn from [42].)

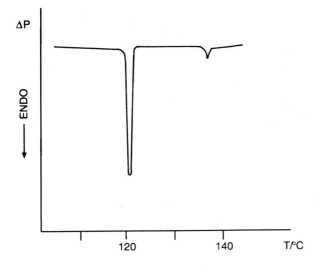

Figure 3.16   DSC curve for *p*-azoxyanisole (15 mg, 16 K/min, static air).

this means that ΔH will be large for a large change in order. Figure 3.16 shows the transformation of *p*-azoxyanisole

$$CH_3O \cdot C_6H_4 \cdot N = NO \cdot C_6H_4 \cdot OCH_3$$

The first transition at 119 °C has a large ΔH because the material goes

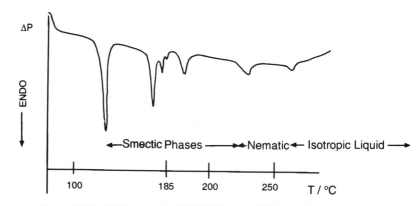

Figure 3.17   DSC curve for OOBPD (9 mg, 10 K/min, flowing nitrogen).

from ordered crystal to disordered, nematic liquid crystal. The second, smaller, transition at 135 °C goes to the true liquid phase.

Even more complex DSC curves are obtained for other systems, as shown in Figure 3.17, for N,N'-bis(4-octyloxybenzylidene)-p-phenylene diamine (OOBPD).

> **Exercise**   By examining the DSC trace of Figure 3.17, suggest probable answers to the following:
>
> 1. What is the most likely phase below 100 °C?
> 2. Given that the phase stable above 205 °C is a *nematic* liquid crystal, suggest the type of phase changes that give rise to the peaks between 120 and 205 °C.
> 3. On cooling this material to 80 °C, after all the liquid crystal phases have formed, *two exothermic* peaks are observed between 110 and 80 °C. What might be happening?

PHASE DIAGRAMS

When more than one chemical component is present, the thermodynamics become even more complicated, and the melting behaviour more complex. The traditional way of studying the phase behaviour of mixtures was by cooling curves, plotting T versus time as a mixture cooled from the melt [46]. DTA or DSC can give more information, more rapidly with smaller samples, provided sensible conditions are used – that is, fairly slow heating rates, preferably less than 10 K/min.

When a pure solid substance A does not dissolve at all in pure solid B, the freezing point of A is lowered by the presence of B, and the freezing point of B is lowered by A, giving the diagram shown in Figure 3.18. The

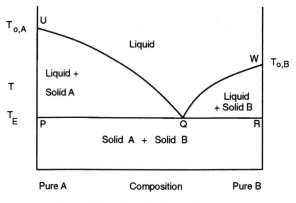

Figure 3.18   Eutectic phase diagram.

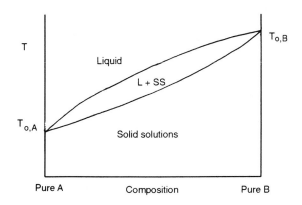

Figure 3.19   Phase diagram for a continuous series of solid solutions.

lowest temperature at which liquid can exist is called the *eutectic temperature* T_E and below this we have various mixtures of crystals of A and crystals of B.

When A and B are very similar in structure, for example, both form cubic crystals, they may form solid solutions, where a mixed crystal structure is obtained, having crystal lattice parameters intermediate between those of A and B. This gives the phase diagram shown in Figure 3.19 [47] or the phase diagram of Figure 3.20.

The DSC or DTA curve of any of the mixtures in these systems generally shows the phase changes.

When the temperature is higher than the solidus line, PQR, the mixture will start to melt and an endothermic peak will be obtained, whose size will depend upon the amount which melts at this temperature. The nearer the mixture is to the eutectic composition Q, the larger the peak at T_E. The

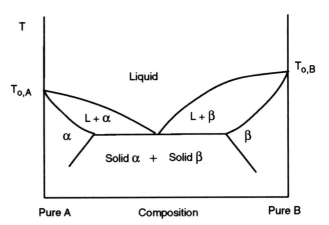

Figure 3.20   Phase diagram for a system with partially miscible solids α and β.

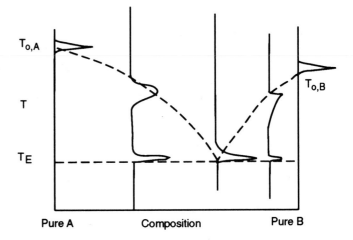

Figure 3.21   Eutectic phase diagram with superimposed DTA curves.

mixture will continue to melt, gradually, until the liquidus line UQW is reached, when the final fraction will become liquid. This temperature is given by the *final* peak of the trace.

Eutectic systems give two peaks, as shown in Figure 3.21. Solid solutions give a single, broad peak, as shown in Figure 3.22. Partially miscible systems give both sorts! When the components react to form compounds, this gives more complex diagrams [47] which may still be studied by thermal methods. The technique has been used extensively to study metallurgical phase diagrams [48,49], mixtures of liquid crystals [50] and pharmaceuticals [40].

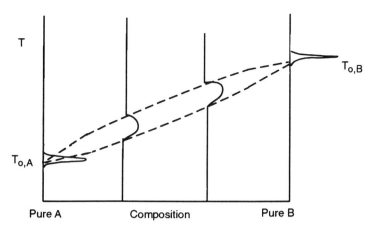

Figure 3.22   Solid solution phase diagram with superimposed DTA curves.

COMPATIBILITY OF PHARMACEUTICAL MIXTURES

This term implies that two or more components of a pharmaceutical dosage form may coexist without interaction during the whole shelf life of the preparation. Often the dosage form contains the active drug component plus other components or excipients to dilute, bind or disperse the drug. If there is physical or chemical interaction, then the physical and pharmaceutical properties of the mixture may be greatly affected. Jacobson and Reier [51] studied stearic acid mixtures with penicillins and showed that DSC methods correlated well with 8 week 50 °C stability tests. If there is no interaction, either physical or chemical, then the dosage mixture should show the same thermal analysis features as the original pure components, in the appropriate proportions. Any changes or new features indicate interaction, as shown in Figure 3.23. Grant *et al.* [52] showed that a mixture of glycerides showed eutectic behaviour, but that this was altered considerably when therapeutic amounts of ketoprofen were added.

Giordano *et al.* [53] showed that although a simple physical mixture of the drug trimethoprim (TMP) with the excipient cross-linked polyvinylpyrrolidone (PVP–XL) showed the same features as the original pure components, grinding the materials together changes the behaviour, as shown in Figure 3.24.

THERMOPLASTIC POLYMER PHASE CHANGES

Polymers which melt before they decompose may crystallise completely, or may remain as irregular, amorphous solids or glasses [54]. On heating a brittle 'glassy' polymer, the change to a more plastic material shows as an increase in heat capacity, as discussed below. The polymer molecules are

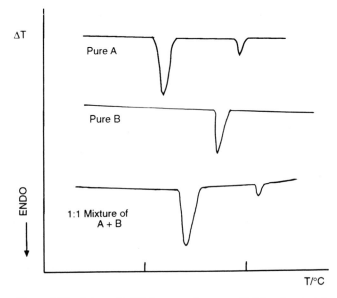

Figure 3.23 Schematic DTA curves for compatibility of materials.

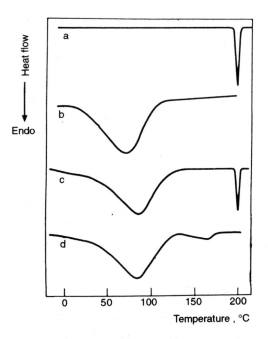

Figure 3.24 DSC traces of (a) PVP–XL (b) TMP, (c) physical mixture of 75% PVP–XL/ 25% TMP, (d) ground mixture of 75% PVP–XL/25% TMP [53].

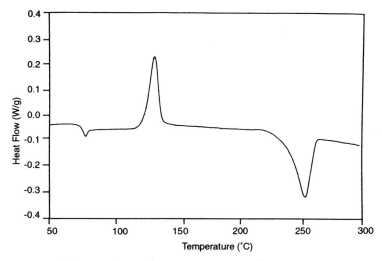

Figure 3.25   DSC curve for poly(ethylene terephthalate), PET, quenched from melt (10 mg, 10 K/min, flowing nitrogen).

then freer to move and can rearrange into the more regular structure of a crystal, giving an exothermic peak of 'cold crystallisation'. Finally, the crystalline polymer will melt to a liquid often over a broad range. With some polymers such as atactic polystyrene, this is difficult to detect. Figure 3.25 shows all these changes for poly(ethylene terephthalate), PET.

Some polymer samples can show multiple melting endotherms, indicating that portions of the sample have crystallised at different temperatures during the previous thermal treatment.

Polymeric samples often do not crystallise completely. The area of the melting peak of a partially crystalline sample may be compared to that of a standard of known crystallinity, or a completely crystalline sample, to give the percentage crystallinity [55].

$$\% \text{ Crystallinity} = \frac{\text{Area of sample melting peak} \times (\% \text{ Standard})}{\text{Area of standard melting peak}}$$

The components in polymer blends may melt separately and give evidence of the materials present and their amounts. Mixtures of poly(propylene) and poly(ethylene) have been analysed by DSC [56] and the components in recycled polymer waste have been detected by DTA [57].

HEAT CAPACITY MEASUREMENTS

The amount of heat needed to raise the temperature of the sample by 1 K is its heat capacity, $C_{p,S}$, in joules per Kelvin and:

$$C_{p,S} = (\partial q_S / \partial T)_p$$

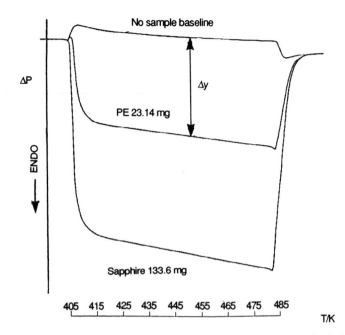

Figure 3.26   DSC curves for heat capacity measurement on molten polyethylene (PE).

On a DSC instrument, the calibrated y axis represents the differential rate of supply of heat energy, or ΔPower or $(d\Delta q/dt)$. Since we are heating at a constant rate, (dT/dt), the product of the y axis deflection Δy from the baseline with no sample, divided by the heating rate, should give the difference in heat capacity due to the sample.

$$C_{p,S} = K \cdot \Delta y \, / \, (dT/dt)$$

where K is a calorimetric sensitivity, found by calibration.

Generally the calibration is done with sapphire, pure, crystalline Al_2O_3, whose heat capacity is well known.

Note: The higher the heating rate, (dT/dt), the greater the deflection obtained!

The 'no sample' baseline can be used to correct for the effects of the reference side.

Exercise   *Given that the heat capacity of sapphire at 445 K is 0.997 J/(K g) and the data in Figure 3.27, calculate C_p for PE. The heating rate is 20 °C/min.*

C_p (sapphire) $= 0.997 \times 133.6 \times 10^{-3} = 0.1332$ J/K
Δy (sapphire) $= 75.0$ mm
C_p (sapphire) $= (K \times 75)/\, 20 = 0.1332$

Thus

$$K = 3.552 \times 10^{-2} \text{ J/(min mm)}$$

For the polyethylene

$$\Delta y = 34 \text{ mm}$$

so

$$C_p(\text{PE}) = (3.552 \times 10^{-2} \times 34)/20 = 0.0604 \text{ J/K}$$

and since the mass of polyethylene is 23.14 mg,

$$C_p(\text{PE}) = 2.609 \text{ J/(K g)}.$$

This agrees quite well with literature values [59]. Wunderlich has published considerable data relating to the heat capacity of polymers [60,61].

The heat capacity is a most important quantity to know for many polymers, insulators and building materials [62].

GLASS TRANSITION TEMPERATURES

Below a certain temperature, known as the glass transition temperature T_g, the polymer segments do not have enough energy to rearrange or to rotate themselves. Such a material is brittle and a *glass*. As the sample is heated, there is a small increase in volume and energy, until at T_g the chains become more mobile and the polymer more plastic or rubbery. Further heating, as shown in Figure 3.27, allows the polymer to crystallise and then melt. Cooling from the melt directly may cause some of the polymer to crystallise at T_m. These changes are shown diagrammatically in Figure 3.27 and the T_g for various polymers is shown in Figure 3.28.

At the glass transition temperature, the heat capacity of the sample increases, since the chains acquire further freedom of movement. Therefore we observe a step and an increase ΔC_p, and also a change in the expansion.

It must be clearly recognised that the T_g is a *time-dependent phenomenon*! A polymer may slowly distort at a low temperature, but may behave in a brittle fashion when bent rapidly. While the T_g bears a marked resemblance to true second-order thermodynamic transitions [64] which always occur over a fixed, equilibrium temperature range, the value obtained for T_g depends greatly on the heating and cooling rates used in the DSC run.

If the heating rate is very low, say 0.1 K/min, the T_g may be low, say 80 °C, and this value is obtained again if the sample is cooled at the same rate. However, if the same sample is heated and cooled at 20 K/min the T_g

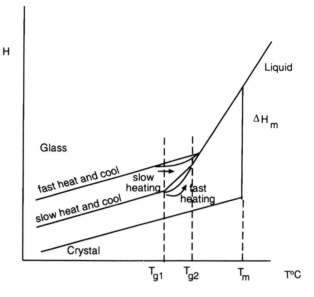

Figure 3.27   Schematic energy changes for material forming a glass.

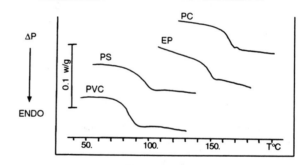

Figure 3.28   DSC curves showing the glass transitions for several polymers. All samples about 20 mg as received, run at 10 K/min in flowing nitrogen. PVC = poly(vinyl chloride); PS = polystyrene; EP = cured epoxy resin; PC = polycarbonate [63].

rises to 85 °C or higher! What happens if the heating and cooling rates are *not* the same? The sample that has been cooled slowly (say 0.1 K/min) goes through its T_g at 80 °C. If it is then heated at 20 K/min it does not transform till 85 °C. This means that the sample must absorb more energy to reach the enthalpy of the rubbery state. This results in an *endotherm*, superimposed on the glass transition step. This is shown in Figure 3.29.

The heating and cooling rates must therefore be stated, and the T_g also depends on molecular weight, degree of cure and amount of plasticiser present [65,66].

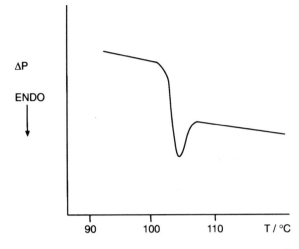

Figure 3.29 DSC curve for polystyrene cooled at 0.2 K/min and then heated at 5 K/min. (After Wunderlich [65].)

PURITY MEASUREMENTS

The determination of the absolute purity of a material is of highest importance for the characterisation of pharmaceuticals, insecticides, pure organics and metals. The theory [67] assumes that there are two effective components that give the eutectic phase diagram of Figure 3.30.

Thermodynamics shows that for a slightly impure A, containing x_B mol% of impurity B, the van't Hoff equation applies and the freezing point is lowered from that $T_{0,A}$ of pure A to T_m [68]. ΔT is given by:

$$\Delta T = (T_{0,A} - T_m) = (RT_{0,A}^2/\Delta H_{fus,A}) \cdot x_B$$

When the mole fraction of impurity is small, ΔT must also be small, but the DSC peaks are found to be much broader for impure samples than for pure. The reason for this involves the *gradual* melting of the impure sample.

Consider the section of the eutectic diagram close to the pure A axis. The lines may be regarded as approximately straight, and we may construct similar triangles as shown. Since melting starts at the eutectic, the tie line PQR may be divided so that the ratio of the liquid solid is

    Fraction liquid/fraction solid = PQ/QR
or Fraction melted, F = PQ/PR

From consideration of similar triangles, this may be put equal to the ratio of temperature intervals on the y axis.

$$F = (T_{0,A} - T)/(T_{0,A} - T_m) = (T_{0,A} - T)/\Delta T$$

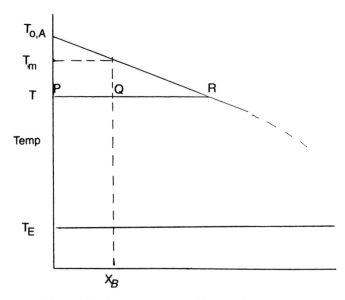

Figure 3.30   Section of eutectic diagram close to pure A.

Substitution into the van't Hoff equation gives:

$$(T_{0,A} - T) = (1/F) \cdot (RT_{0,A}^2/\Delta H_{\text{fus,A}}) \cdot x_B$$

or, rearranging,

$$T = T_{0,A} - (1/F) \cdot (RT_{0,A}^2/\Delta H_{\text{fus,A}}) \cdot x_B$$

This predicts that a plot of temperature versus $(1/F)$ should be a straight line of slope dependent on x_B, $T_{0,A}$, and $\Delta H_{\text{fus,A}}$. The temperature T may be corrected for thermal lag, measured using the peak for the melting of pure indium (Figure 3.11) or by the built-in software of the instrument [69].

From a DSC trace (Figure 3.31) obtained at the slowest heating rate possible, we may measure $\Delta H_{\text{fus,A}}$ and $T_{0,A}$ approximately, and the fraction melted at T °C from the ratio of the area a up to T, divided by the total peak area A.

$$F = a/A$$

Ideally this gives a good straight line. However, there are several factors which combine to produce a curve:

- neglect of any pre-melting before the peak measured
- non-equilibrium conditions, due to dynamic heating
- solid-solution formation.

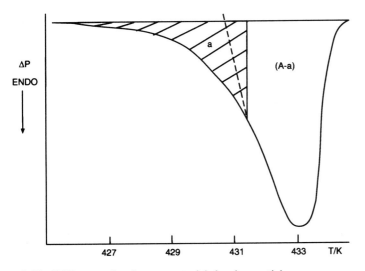

Figure 3.31   DSC trace of an impure material showing partial area measurement and temperature correction.

A correction for these factors that has been shown to be effective involves the use of an add-on correction da

$$F' = (a+\mathrm{d}a)/(A+\mathrm{d}a)$$

and

$$\Delta H_{\mathrm{fus,A}} = K\cdot(A+\mathrm{d}a)$$

This can be related to the modified van't Hoff equation for solid solutions [70]

$$\Delta T' = (1-k)\cdot(RT_{0,\mathrm{A}}^2/\Delta H_{\mathrm{fus,A}})\cdot x_{\mathrm{B}}$$

where k is the distribution coefficient for the solid solution.

In a different approach to eliminating the non-equilibrium effects, the use of discrete steps in heating has been suggested. If a step of less than 1 K is used and the system is then allowed to equilibrate, the trace produced shows peaks of equal area until melting commences. The peaks then increase until melting is complete. The sum of the peak areas, corrected for the heat capacity effect, gives values for a and A.

Plato and Glasgow [71] used the method to measure the purity of many compounds and it forms the basis of the ASTM method E928–85. Comparing the DSC and NMR methods of determining purity, Garn *et al.* [72] have shown that solid-solution formation plays a large role in determining the melting behaviour.

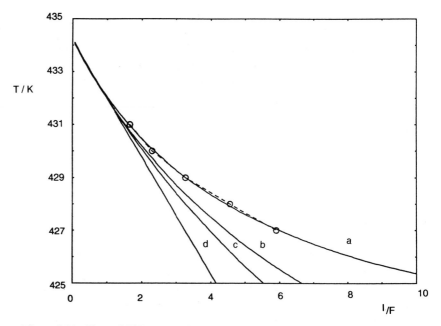

Figure 3.32 Plots of T/K versus $1/F$ and corrected $1/F$: (a) uncorrected; (b) 5% correction; (c) 10% correction; (d) 19% correction.

Example *A 3 mg sample of an insecticide (RMM = 365) was melted on a DSC at 1 K/min and gave the following results:*

T/K	a/cm^2	F	Total area, $A = 10.5\ cm^2$
427	1.78	0.170	$K = 10.0\ mJ/cm^2$
428	2.31	0.220	
429	3.21	0.306	
430	4.58	0.436	
431	6.33	0.603	

Calculate $T_{0,A}$, $\Delta H_{fus,A}$ and x_B.

A straightforward plot of T against $1/F$ gives a curve, intersecting the T axis at about 434 K (Figure 3.32(a))

We may treat the data in three ways:

1. Continue adding small amounts to a and A until a straight line is obtained;
2. Plot $(1/(T_{0,A} - T))$ versus F. The slope should give k.
3. Do a regression analysis of the curve.

1. Successive additions of 0.5 cm² to the area give a best line when the correction is about 2.0 cm², and this gives a a corrected total area of 12.5 cm², so that the corrected ΔH is about 15.2 kJ/mol and the slope of -2.2 gives an x_B value of 2.1 mol% (97.9% pure A). This is shown in Figure 3.32.

2. If we rearrange the equations above, we find that a plot of $1/(T_{0,A} - T)$ versus F should be a straight line of slope $(\Delta H/RT^2_{0,A})/x_B$ and intercept related to the correction, k. If we assume that $T_{0,A} = 434$ K and make this plot, we get a value of x_B of 0.028, or a purity of 97.2%.

3. Although this requires considerable time, or computation, it is probably the best way ([19], Ch. 14). Regression analysis gives a correction of 0.163 and a purity of 97.9%.

3.9.2 *Chemical reactions*

KINETICS OF REACTION FROM DSC TRACES

Chemical reactions, and some physical changes such as cold crystallisation, occur at a temperature-dependent rate and involve an energy change, ΔH. We have seen how this may be related theoretically to the peak area and we may make assumptions about the relationship between the rate of supply of heat, which gives the ΔP signal, and the rate of reaction. For the heat capacity part of the change, which does *not* depend on the rate of reaction but on the material present at a particular time, we can draw in a proposed baseline. We must also use small samples and control the instrument carefully so that the temperature gradients within the sample are very small.

Provided we may make these assumptions, the peak that we obtain, either for a scanning or for an isothermal run, may be divided into fractions representing the fraction of reaction α which has occurred.

The kinetic equations have been described in Chapter 2, and may be very complex. The analysis of data is best done for a simple single-stage process. If we are dealing with a multi-stage reaction, for example the decomposition of a polymer, it is very difficult to assign kinetic parameters to the stages of reaction. Additional information is needed for the separate values of ΔH for each stage. The separate rates might be obtained by combining DSC with other analyses of products, for example by infrared spectrometry or evolved gas analysis.

Barrett [73] applied this method to the decomposition of free-radical initiators like azobisisobutyronitrile (AIBN). The DSC of AIBN in dibutyl phthalate at heating rates between 4 and 32 K/min gave kinetic data corresponding well to first order kinetics and an activation energy of about 125 kJ/mol in agreement with UV experiments.

Crystallisation rates of polymers have been measured [74, 75] using the partial area to determine the percentage crystallized. The results often fit an Avrami equation:

$$[-\ln(1-\alpha)]^{1/n} = kt$$

where α is the fraction of the final crystallinity developed at time t, and n

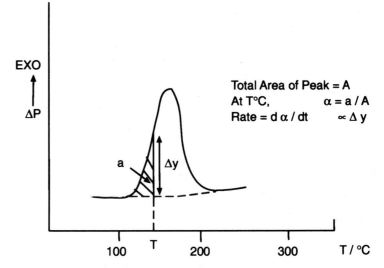

Figure 3.33 DSC curve for exothermic reaction showing measurement of partial and total areas.

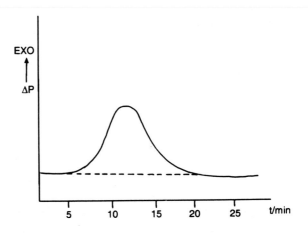

Figure 3.34 Isothermal DSC curves for polymer crystallisation.

depends on the mechanism of crystal growth, e.g. $n = 3$ for spherical growth.

The assessment of thermal hazards by analysing kinetic data is a standard ASTM method E698-79 [76]. The method recommended by this committee is that of Ozawa [77], where samples are run at several different heating rates, β, the temperature corresponding to the peak maximum T_{max} is noted. Ozawa showed that if $\ln(\beta)$ is plotted versus $1/T_{max}$ a straight line of slope approximately E/R is obtained, and the values may be refined by using a correction factor.

3.9.3 *Inorganic compounds and complexes*

The dehydration, decomposition and other reactions that take place with inorganic chemicals and minerals were among the first reactions studied by DTA. We may obtain vital information about the endothermic or exo-thermic nature of the reactions, which may be compared with data from thermodynamic tables. We can detect melting and other phase changes, and the DTA (or DSC) trace itself is an effective 'fingerprint' of the thermal behaviour of the compound *under those particular conditions!*

CALCIUM OXALATE MONOHYDRATE $(CaC_2O_4 \cdot H_2O)$

Figure 3.35 shows the DTA curves of calcium oxalate monohydrate run under flowing air and under flowing nitrogen. The endothermic peak around 200 °C is due to the loss of the hydrate water, and is similar in both gases. In nitrogen, the peak around 400 °C is also endothermic, and corresponds to the breakdown of the oxalate with loss of carbon monoxide. When the atmosphere is oxidising, however, the CO is oxidised as soon as it is formed, and the peak becomes exothermic. The endothermic peak at 800 °C is due to the decomposition of the calcium carbonate, and is only slightly affected by the change of gas.

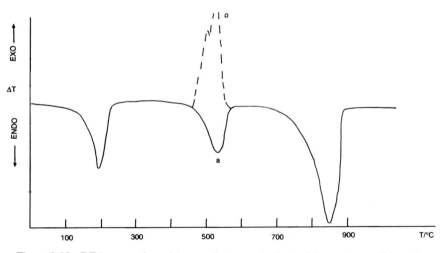

Figure 3.35   DTA curves for calcium oxalate monohydrate: (a) in nitrogen; (b) in air. Sample: 10 mg of powder, heated at 10 K/min.

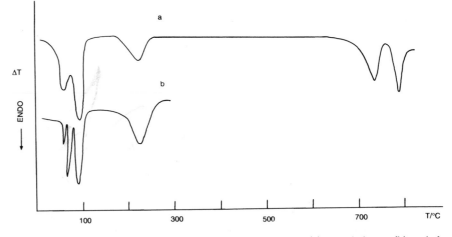

Figure 3.36   DTA curves for copper sulphate pentahydrate: (a) unsealed pan; (b) sealed pan with pinhole. Sample: 6 mg crystalline powder, 10 K/min, flowing air.

The effects on the DTA curve of changes in heating rate, sample mass and particle size, atmosphere and gas flow rate have all been thoroughly investigated for this system [78].

COPPER SULPHATE PENTAHYDRATE ($CuSO_4 \cdot 5H_2O$)

The dehydrations and decomposition of copper sulphate produce a series of endothermic peaks on the DTA or DSC trace (Figure 3.36).

The initial loss of water may produce traces that seem different, especially if different sample pans are used. A very open, flat pan will give two broad, overlapping peaks of roughly equal area around 120 °C. If the loss of water is restricted – for example, by using a lidded pan with a small pinhole only – then the water may be retained and boil off separately, giving a triple peak. At higher temperature, the final molecule of water is lost.

Calorimetric measurements on the dehydration peaks gives ΔH values of approximately:

$$CuSO_4 \cdot 5H_2O = CuSO_4 \cdot 3H_2O + 2H_2O \quad \Delta H \text{ (373 K)} = 100 \text{ kJ/mol}$$
$$CuSO_4 \cdot 3H_2O = CuSO_4 \cdot H_2O + 2H_2O \quad \Delta H \text{ (400 K)} = 104 \text{ kJ/mol}$$
$$CuSO_4 \cdot H_2O = CuSO_4 + H_2O \quad \Delta H \text{ (510 K)} = 72 \text{ kJ/mol}$$

The sample decomposes to the oxide at higher temperatures:

$$CuSO_4 = CuO + SO_2 + \tfrac{1}{2}O_2$$

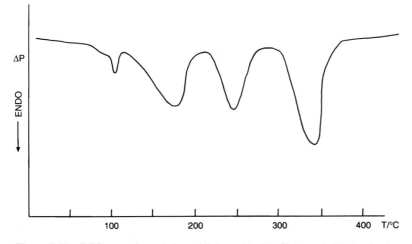

Figure 3.37   DSC curve for cobalt pyridinium chloride (3.4 mg, 8 K/min, flowing nitrogen) [81].

METAL COMPLEXES

Very many complex compounds have been investigated by DSC [79, 80, 81 and general texts] and the effects of the metal and the ligand on the reactions and the reaction enthalpies measured. The important factors are the loss of the ligand, phase changes and decompositions of the sample and of the products. For these complicated reactions, it is advantageous to use *simultaneous techniques* which will be discussed in Chapter 5. One example of the calorimetric use of DSC is cobalt pyridinium chloride, $Co(C_5H_5N)_2 \cdot Cl_2$ [81], which gives four transitions up to 500 °C. At about 120 °C, a small, sharp endothermic peak with $\Delta H = 12.6$ kJ/mol is obtained, but this corresponds to no mass loss in the TG experiment. Three subsequent peaks at 180, 280 and 350 °C, show progressive losses on the thermobalance, corresponding to loss of 1, 1/3, and 2/3 pyridine molecules respectively. The total ΔH for these reactions is about 120 kJ/mol(Fig. 3.37).

HIGH ALUMINA CEMENTS [82–85]

Ordinary Portland cement (OPC) is made by heating clays, which are aluminosilicates, with calcium carbonate, and consists chiefly of mixed calcium silicates. During their use, the cement is mixed with water which hydrates the silicates and produces a good cementing material, which very slowly develops its final strength over many months. The DTA curve of an OPC is given in Figure 3.38, which shows the dehydration of the material and also the peak due to the quartz transition at 573 °C.

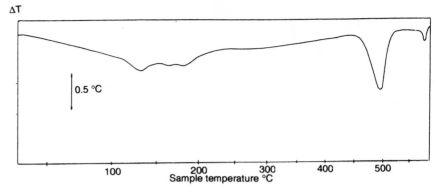

Figure 3.38 DTA curve for a Portland cement concrete sample (50 mg, 20 K/min, nitrogen).

High alumina cement (HAC) is made from bauxite and limestone, and is mostly a mixture of calcium aluminates. These react with water and gain strength more rapidly than OPC, and so they speed up the building process. Unfortunately, under certain conditions, beams constructed of HAC concrete collapsed catastrophically. A programme of testing was undertaken at once in which DTA played a most significant part.

The material which is first produced in the hardening process is the decahydrate

$$CaO \cdot Al_2O_3 \cdot 10H_2O \quad \text{abbreviated to } CAH_{10}$$

This is not the most stable hydrate, and may gradually undergo a 'conversion' reaction to more stable compounds, such as the hexahydrate and hydrated alumina or gibbsite.

$$3CaO \cdot Al_2O_3 \cdot 6H_2O \text{ (or } C_3AH_6) \quad \text{and} \quad Al_2O_3 \cdot 3H_2O \text{ (or } AH_3)$$
$$3(CaO \cdot Al_2O_3 \cdot 10H_2O) = 3CaO \cdot Al_2O_3 \cdot 6H_2O + Al_2O_3 \cdot 3H_2O + 18H_2O$$

This conversion weakens the structure, because the products of reaction are more dense and the reacted structure more porous, particularly if conversion has taken place rapidly.

The decahydrate, the hexahydrate and gibbsite, plus several other hydrates, all lose water over specific temperature ranges, as indicated in Figure 3.39. All the dehydrations are endothermic, and the sample will show these dehydrations clearly by DTA. It is also helpful to run to 600 °C to show the silica peak, but also to show the dehydration peak of $Ca(OH)_2$ at 500 °C which is characteristic of OPC.

The method of sampling is important, since high-speed drilling might heat the sample above the dehydration temperatures. Plaster should be removed, since the dehydration peaks of gypsum and plaster of Paris occur in the region of interest for HAC. The slowly drilled out samples should be

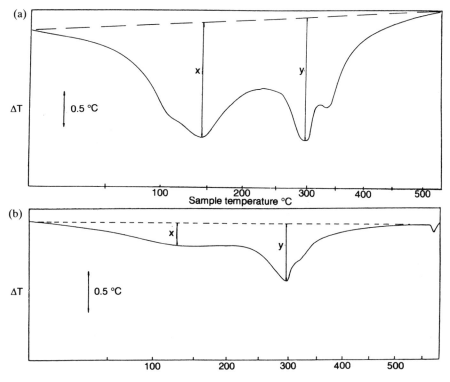

Figure 3.39 DTA curves for HAC standards having conversions of (a) 50% and (b) 70% [83].

sieved to remove large particles, and metallic particles from the drilling should be removed with a magnet.

Samples of 10–100 mg are run in flowing nitrogen at heating rates between 10 and 30 K/min, and calibration standards run under the same conditions. The 'degree of conversion', D_c, is calculated from the *height* of the peaks measured, as shown on Figure 3.39.

$$D_c = \frac{100 \cdot \text{Amount of AH}_3}{(\text{Amount of AH}_3 + \text{amount of CAH}_{10})}$$

$$\text{or} \quad D_c = \frac{100 \cdot a \cdot (\text{Height of AH}_3 \text{ peak})}{(a \cdot \text{Height of AH}_3 \text{ peak} + b \cdot \text{Height of CAH}_{10} \text{ peak})}$$

$$\text{or} \quad D_c = \frac{100 \cdot (\text{Height of AH}_3 \text{ peak})}{(\text{Height of AH}_3 \text{ peak} + K \cdot \text{Height of CAH}_{10} \text{ peak})}$$

where a and b are calibration constants and $K = b/a$.

Examples of the calculation of degree of conversion should not be quoted to better than ± 5%

Calculation   The calibration constant K was determined by running standard samples with 50 and 70% conversion. For the 50%, the AH_3 peak height was 3.5 cm and the CAH_{10} peak height 3.7 cm. Therefore:

$$50\% = 100 \times 3.5/(3.5 + K \times 3.7)$$

Thus $K = 0.95$.

For an unknown HAC sample, the AH_3 peak was 4.4 cm high, and the CAH_{10} peak 2.6 cm high. Therefore:

$$D_c = 100 \times 4.4/(4.4 + 0.95 \times 2.6)$$
$$D_c = 64 \pm 5\% \text{ conversion.}$$

CLAYS AND OTHER MINERALS

The literature describing applications of DTA on minerals is very large [82, 86–88], and only a few typical examples will be described here.

We have already seen from the HAC experiments that minerals give characteristic DTA traces. Single component minerals, such as quartz, show their phase transitions. Hydrated and hydroxyl minerals show dehydration peaks, and carbonate minerals lose carbon dioxide. Typical traces are shown in Figure 3.40.

Kaolinite, or china clay, is an important industrial mineral. The DTA trace of Figure 3.41 shows a small, broad endothermic peak at about 100 °C and a much larger endotherm at 550 °C. A sharp *exotherm* appears at about 1000 °C. These have been interpreted as loss of moisture at 100 °C, dehydroxylation by 700 °C and the reaction to form a new crystalline material, chiefly mullite ($3Al_2O_3 \cdot 2SiO_2$).

Borate minerals have very complex structures and cannot always be characterised by elemental analysis. For a series of minerals of general formula $M_2B_6O_{11} \cdot xH_2O$, some containing discrete anions (Figure 3.42(a)) lose most of their water below 250 °C while others which have linked chains do not decompose until much higher temperatures (Figure 3.42(b)). Mixtures of these two types may be analysed semi-quantitatively by thermal methods [91].

3.9.4   *Synthesis of compounds at high temperatures*

The high-temperature reactions of inorganic compounds produce many new materials, some of which are of special interest to the electronics industry.

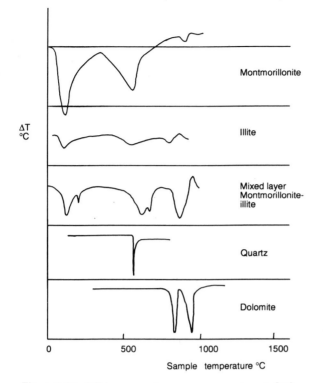

Figure 3.40 DTA curves of some common minerals [89].

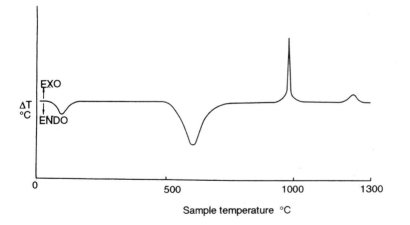

Figure 3.41 DTA curve of kaolinite [89].

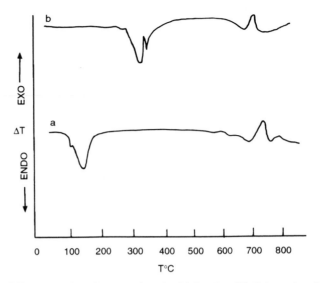

Figure 3.42 DTA curves for triborate minerals: (a) Inyoite; (b) Colemanite. Samples of about 20 mg, 10 K/min, static air [91].

Heating together barium carbonate and iron(III) oxide in the molar ratio 1:6 gives the DTA shown in Figure 3.43.

When heated alone, barium carbonate shows sharp crystal phase transitions at 810 and 970 °C while iron oxide shows a transition at 675 °C. Together, the mixture still shows the transition at 675 °C but now a large, broad endotherm appears with a maximum at 870 °C. X-ray analysis of the final product shows it to be barium hexaferrite, $BaFe_{12}O_{19}$. This material is a valuable magnetic solid used as cores for inductive components.

Similar experiments can be conducted to produce a phosphor from barium hydrogen phosphate $BaHPO_4$ heated with about 10% TiO_2 to 1070 °C, when a material is produced which phosphoresces brightly in the visible when illuminated with UV light [93].

3.9.5 *Pyrotechnics*

The reactions involved in pyrotechnics may be studied by DTA or DSC since the small samples and programming allow good control of the system. The reactions of 'black powder' (powdered sulphur, potassium nitrate and charcoal) have been studied by DTA, TG and DSC [94]. The DTA in nitrogen shows the phase transitions of sulphur and of KNO_3, vaporisation of sulphur followed by the exothermic reaction between charcoal and potassium nitrate between 390 and 550 °C and the decomposition of unreacted potassium nitrate (Figure 3.44). In air, there are two exotherms

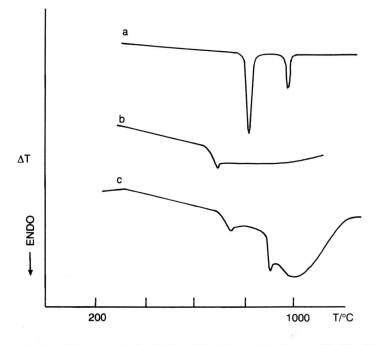

Figure 3.43   DTA curves for (a) $BaCO_3$, (b) α-Fe_2O_3, (c) mixture of $BaCO_3$:$6Fe_2O_3$.

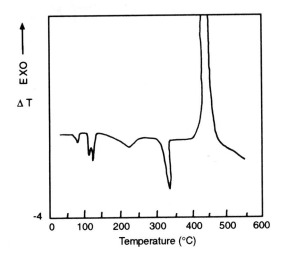

Figure 3.44   DTA of black powder in nitrogen at 20 K/min (After [94]).

due to the oxidation of sulphur at above 190 °C and then oxidation of charcoal. A review of applications of thermal analysis to pyrotechnic systems gives many other examples [95].

3.9.6 *Superconductors*

The discovery of the new materials whose superconducting properties extended to much higher temperatures than before, has given a great impetus to studies of high T_c superconducting ceramics. The synthesis of yttrium barium copper oxide materials has been reviewed by Ozawa [96], and Chen and Sharp report DTA studies on a high-temperature synthetic method [97].

The reaction between Y_2O_3, BaO_2 and Cu metal in the proportions 0.5:2:3 on heating in oxygen is exothermic, as shown in Figure 3.45. They showed that there is no reaction between Y_2O_3 and copper when they are heated alone, but that Y_2O_3 and BaO_2 reacted exothermically to produce Y_2BaO_4. Copper and barium peroxide reacted very vigorously to produce $BaCuO_2$, and the ternary mixture gave a DTA which resembled a combination of these two reactions, producing $YBa_2Cu_3O_y$ ($y = 6–7$) and Y_2BaCuO_5 detected by X-ray diffraction.

3.9.7 *Organic compounds*

The reactions of organic materials and of polymeric materials, both natural and synthetic, have been widely studied by DTA and DSC. The effects of the atmosphere are very important here, since oxidative degradation may occur by a very different mechanism to non-oxidative changes.

OXIDATIVE DEGRADATION

Polymers and also oils degrade when heated in oxidising atmospheres. A standard test for this has been devised and may be used to compare the stabilities of polyolefins and of other oils and fats.

The sample may be heated in nitrogen to 200 °C and the atmosphere then changed to oxygen, with the temperature held constant at 200 °C. The time for the onset of the exothermic oxidation is then noted. Alternatively, the polymer may be heated in oxygen and the temperature at which the onset of oxidation occurs noted.

The same type of test may be applied to lubricants, and to edible oils and fats, and has been used to test the efficiency of anti-oxidants.

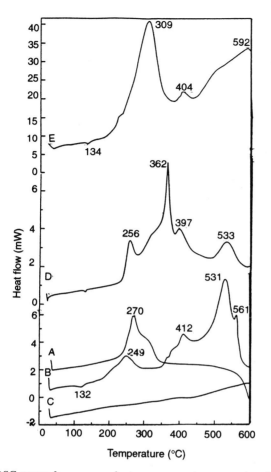

Figure 3.45   DSC curves for superconductor components: curves A to D 10 K/min in
nitrogen; curve E 50 K/min in nitrogen [95].
A: $0.5Y_2O_3 + 2BaO_2$ B: $2BaO_2 + 3Cu$
C: $0.5Y_2O_3 + 3Cu$ D and E: $0.5Y_2O_3 + 2BaO_2 + 3Cu$

POLYMER CURE [100]

The reaction of small molecules to produce larger molecules with different
properties and more stability is generally exothermic. The average ΔH of
polymerisation of unsaturated molecules like styrene and vinyl chloride is
about -100 kJ/mol. The reactions of thermosetting polymers, such as
epoxy resins and polyesters cross-linked with styrene, are also exothermic
and readily studied by DSC.

The reaction of the epoxy group with a curing agent, often an amine, is
the initial stage in the reaction and is followed by a further reaction of the
secondary amine with a further epoxy group.

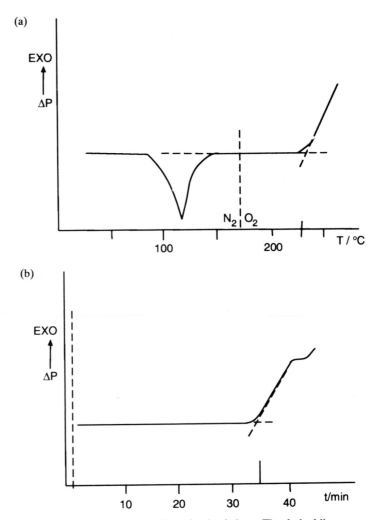

Figure 3.46 DSC curves for oxidation of polyethylene. The dashed line represents the changeover from nitrogen to oxygen. (a) Scanning DSC for PE film, 10 K/min; onset temperature of oxidation = 220 °C. (b) Isothermal DSC for PE film at 200 °C; onset time = 35 min.

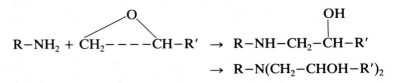

This reaction occurs even at low temperatures, but is rapid above 100 °C,

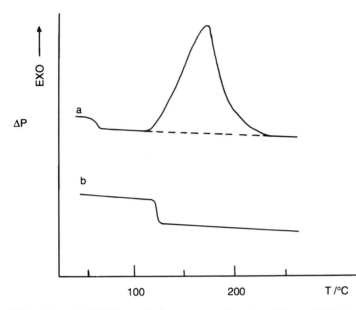

Figure 3.47   Schematic DSC for typical epoxy cure. Samples of 5 mg, 10 K/min, nitrogen. (a) Initial run of uncured sample; (b) re-run of fully cured material.

giving the DSC trace of Figure 3.47. The initial T_g of the starting material is followed by the large exotherm of the reaction. Re-running the cured material shows a new T_g at a much higher temperature. Post-curing will raise this T_g to even higher temperatures.

The kinetics of the reaction may be studied from scanning or an isothermal experiment.

PROTEIN DENATURATION

Proteins may be classified as fibrous, with long thread-like molecules, for example collagen, or globular, with compact 'spherical' shapes such as insulin. Both types have well-developed structures involving folds, coils or sheets and even a helix within a helix: a 'super-helix'. These structures are destroyed by heating or denaturation at extreme pH, and this reaction is endothermic.

Figure 3.48(a) shows the denaturation of collagen, a single material, while Figure 3.48(b) shows the complex behaviour of multi-component systems found in real animal muscle proteins. Note the large sample and slow heating rate used here [102].

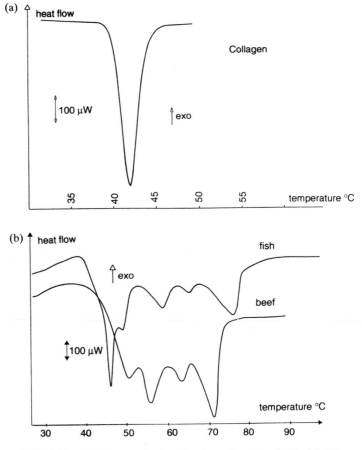

Figure 3.48 Micro-DSC curves for denaturation of proteins [102]. (a) 800 mg, 0.3% collagen solution, sealed vessel, 0.5 K/min; (b) 850 mg beef and fish muscle, sealed vessel, 0.2 K/min. (Courtesy SETARAM)

POLYMER DEGRADATION [20,103]

The complex reactions that occur when polymers degrade in inert or oxidative atmospheres have been discussed in Chapter 2. DTA and DSC studies will give further information on the stages and enthalpy changes involved.

The DTA curves for the oxidative decomposition of poly(vinyl chloride) (PVC) and of polypropylene (PP) are shown in Figures 3.49 and 3.50.

PVC powder shows a small glass transition around 80 °C and then a small endotherm near 300 °C, almost immediately followed by a very large exotherm peaking at around 550 °C. These later stages correspond to the

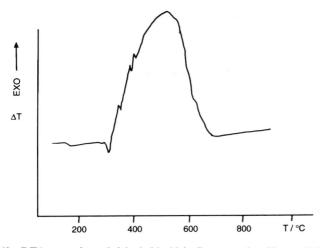

Figure 3.49   DTA curve for poly(vinyl chloride). Geon powder, 20 mg, 15 K/min, flowing air.

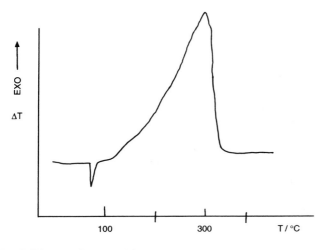

Figure 3.50   DTA curve for unstabilised polypropylene film, 20 mg, 15 K/min, flowing air.

degradation with loss of HCl and flammable volatiles, followed by oxidation of the char which is formed.

PP decomposes in a single stage and the products are then easily oxidised, so a single endothermic peak is obtained for the melt, followed by a large oxidation peak.

3.10
Specialist DSC systems

3.10.1 *Pressure DSC (PDSC)*

By placing the entire DSC or DTA cell in a pressure-tight enclosure, it is possible to work at pressures less than atmospheric, or more than atmospheric.

A simple modification of DSC and DTA equipment [104] allowed reasonably accurate determination of boiling points of organic liquids in the pressure range 20–760 mm Hg. Seyler [105] investigated the parameters affecting this type of measurement, especially the cell design and sample size. Recent work [106] using hermetically sealed pans with 50–100 μm laser-drilled holes gives vapour pressures within a few percent. Results for the vaporisation of water at pressures up to 3.5 MPa gave excellent agreement with literature data [107].

It should be noted that in *raising* the boiling point, or sublimation temperature, any vaporisation which interferes with melting is suppressed, and better melting point, purity and phase data will be obtained.

In predicting the oxidative stability of oils, polymers and foodstuffs, the use of higher pressures of oxygen decreases the volatility of the material (because the boiling point increases with pressure increase) and also increases the concentration of reactant gas. Shorter test times are then possible, and if any volatiles are lost at high temperatures, then lower test temperatures and higher oxygen pressures may be used. These advantages are discussed by Thomas [108]. Oxidation and ignition of oils under a high pressure of nitrogen (8 MPa) with flowing oxygen was used to compare new and used oil samples and anti-oxidants [107].

Reductions of organic compounds with hydrogen using metal catalysts has been studied with PDSC. With platinum or palladium metal catalysts on a silica substrate, the catalyst sample is heated to reaction temperature in helium under pressure, and then the gas switched to hydrogen. Chemisorption and catalytic reduction produce an exotherm on the DSC, which can be related to the hydrogen consumed. If the desired ratio of reactant to react with 2–3 mg of catalyst is put into the DSC cell, a similar procedure produces an exotherm for the reduction of the reactant, for example, *m*-dinitrobenzene using a 5% Pd/carbon catalyst, as shown in Figure 3.51 [109].

Phase diagrams and reactions for praseodymium oxides, reduction of metal oxides, formation of a metal hydride and oxidation of charcoal [107] all demonstrate the usefulness of a pressure DSC system.

3.10.2 *Photocalorimetric DSC*

Polymerisation may be initiated by ultraviolet light, and this is used in many disciplines such as electronics, coatings and dental work. DSC cells have been adapted to measure the polymerisation exotherm while the

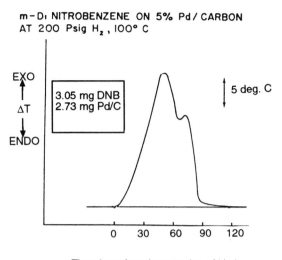

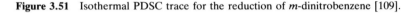

Time (sec. from introduction of H_2)

Figure 3.51   Isothermal PDSC trace for the reduction of *m*-dinitrobenzene [109].

sample is irradiated with UV [110]. The 'photocalorimeter' is made up of a DSC and an accessory which illuminates the sample cell with UV light of selected wavelength and intensity.

Tryson and Schultz [111] investigated the polymerisation of lauryl acrylate and of 1,6-hexanediol diacrylate using a DSC illuminated by a medium-pressure mercury lamp (λ_{max} = 365 nm) through a heat filter filled with de-ionised water and a manually operated shutter. The pans were modified so that a uniform sample thickness was maintained. A typical DSC cure curve is shown in Figure 3.52.

Manley and Scurr [112] used a DTA apparatus to study the UV curing of surface coatings of methyl methacrylate plus azobisisobutyronitrile (AIBN) and Hodd and Menon [113] investigated various initiators for the photopolymerisation of acrylates.

3.10.3   *Modulated DSC (MDSC^{TM})*

The many changes that can be detected by DSC are both an advantage and a disadvantage! If the primary purpose is to detect a glass transition, and a second event such as a chemical reaction or crystallisation overlaps, how may we separate them?

In a conventional DSC, a constant, linear heating rate is applied. If the heating is 'modulated' by a small alternating amount of power supplied in addition to the normal programmed heating, the temperature should follow the profile shown in Figure 3.53 [114,115].

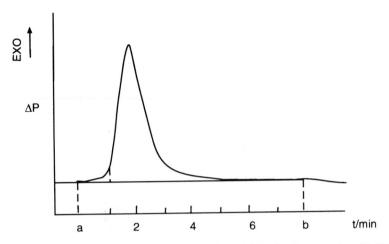

Figure 3.52 Schematic photo–DSC curve for a photo-initiated polymerisation. (a) UV on; (b) UV off.

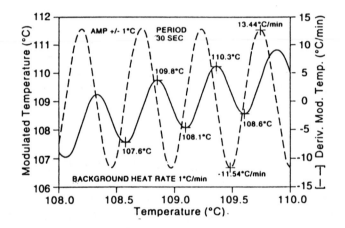

Figure 3.53 MDSC heating profile showing the overlaid ripple on the heating ramp.

When we have a heating programme such that

$$T = T_0 + \beta t + B \sin(\omega t)$$

where ω is the angular frequency ($= 2\pi f$), T_0 is the starting temperature and B is the amplitude of temperature excursion, then the heat flow is given by:

$$(dq/dt) = C_p[\beta + B \cdot \omega \cos(\omega t)] + \bar{f}(t,T) + C \sin(\omega t)$$

where C_p is the heat capacity, $\bar{f}(t,T)$ is the average underlying kinetic

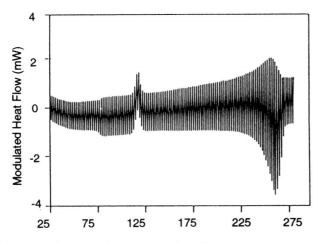

Figure 3.54 MDSC trace for quenched PET before deconvolution.

response and C is the amplitude of the kinetic response to the sine wave modulation.

The heat flow has a cyclic component which is made up of the sine and cosine terms, the amplitude of which is determined by a Fourier transform analysis. The underlying signal, which is equivalent to a conventional DSC, is calculated by an averaging process. The value of C is often low, so that the third term is negligible and the cyclic signal gives the heat capacity directly. By multiplying this by the heating rate the *reversing heat flow* is obtained. This is then subtracted from the underlying signal to obtain the *non-reversing heat flow*.

The main component of the reversing signal is the heat capacity, and thus the glass transition appears in this signal.

Other events which appear in the non-reversing signal are decomposition, curing of polymers, cold crystallisation, and molecular relaxation at T_g. The total signal is shown in Figure 3.54 and the signals separated by a discrete Fourier transform into the reversing and non-reversing parts are shown in Figure 3.55.

The reversing part shows the glass transition and the melting as the main events. The non-reversing shows the relaxation occurring at T_g plus the cold crystallisation and the non-reversing component at T_m.

Because we may separate the events, each may be seen more clearly and measured more accurately. Heat capacity may be measured directly, and T_g separated from reactions occurring over the same temperature range, as shown in Figure 3.56.

Applications to reversible and metastable transitions of liquid crystals, or pharmaceuticals and polymers have been reported [116,117].

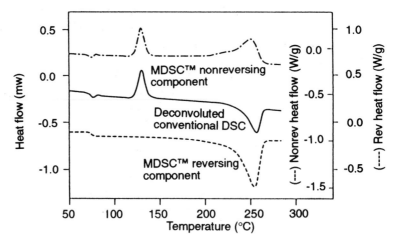

Figure 3.55   Resolved components from MDSC of PET showing the standard heat flow, reversing and non-reversing heat flows.

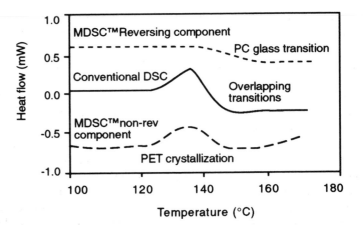

Figure 3.56   MDSC curves for a bilayer film of polycarbonate (PC) and PET. The overlap of the T_g of PC and the crystallisation exotherm of PET is resolved.

3.11
Problems
(Solutions on p. 275)

1. Which of the following changes could NOT be detected by DTA: (a) loss of moisture; (b) sublimation; (c) desorption of vapour; (d) polymer surface softening?

2. When the heating rate is changed on a DSC or DTA instrument, without changing the sensitivity, what will be the effect on: (a) the baseline; (b) any glass transition step; (c) an endothermic melting peak; (d) an exothermic reaction peak?

3. Pure naphthalene melts at 80 °C. Pure 1-naphthol melts at 123 °C. All mixtures of these two components show a *single* peak when heated and a 50 mol% mixture shows a single broad peak around 100 °C. What type of phase diagram does this suggest?

4. Pure phenacetin melts at 134.6 °C. From the DSC its heat of fusion is 53.3 kJ/mol. If the *corrected* fractions melted are 0.20 at 133.0 °C and 0.29 at 133.5 °C, estimate the purity of the phenacetin.

5. For the high-temperature reaction of copper sulphate:

$$CuSO_4 \rightarrow CuO + SO_2 + \tfrac{1}{2}O_2$$

given the data below, estimate the ΔH for this reaction and comment on the value with respect to the DTA trace given in Figure 3.36.

Compound	ΔH_f/kJ/mol at 298 K
$CuSO_4$(s)	−800
CuO(s)	−157
SO_2(g)	−297
O_2(g)	0

This estimate will be rather incorrect! Why?

6. A series of DSC runs were done on an epoxy coating material. The first, run up to 100 °C only, showed a small endothermic step at 50 °C. The second, run up to 150 °C, showed a large exothermic peak around 140 °C. The third, re-running the second material up to 200 °C, showed a smaller exothermic peak around 140 °C. The fourth showed *no* exotherm, but only an endothermic step at about 100 °C. Please explain!

7. *Without using a thermobalance*, how would you distinguish between DTA or DSC peaks that are due to structural inversions and those that are due to chemical reactions with loss of mass?

8. A mineral felspar is reported to be suspect although its chemical breakdown into Al_2O_3 and SiO_2 reveals an ideal felspar. A DTA curve shows three peaks: endothermic peaks at 600 °C and 750 °C and an exothermic peak at 990 °C. If the first and last peaks are characteristic of kaolin, $Al_2O_3 \cdot 2SiO_2 \cdot 2H_2O$, suggesting the felspar is contaminated, how would you estimate the amount of kaolin in the suspect material, given a 'pure' felspar which shows a peak at 750 °C and a pure kaolin sample?

References

1. A.L. Lavoisier, P.S. de Laplace, *Mem. R. Acad. Sci.*, *Paris*, 1784, 355.
2. J.B. Fourier, *Theorie Analytique de la Chaleur*, Paris, 1822.
3. J.P. Joule, *Collected Works*, The Physical Society, London, 1884, p. 474.
4. J.P. McCullogh, D.W. Scott, *Experimental Thermodynamics*, Vol. 1, Butterworth, London, 1968.
5. H. Le Chatelier, *C.R. Acad. Sci.*, *Paris*, 1887, **104**, 1443.
6. W.C. Roberts-Austen, *Proc. R. Instn. Mech. Eng.* London, 1899, p. 35.
7. E.S. Watson, M.J. O'Neill, J.Justin, N.Brenner, *Anal. Chem.*, 1964, **36**, 1233.
8. J.O. Hill, *For Better Thermal Analysis and Calorimetry* (3rd edn), ICTAC, 1991.

9. R.C. Mackenzie, *Anal. Proc.*, 1980, 217.
10. R.L. Blaine, *A Generic Definition of Differential Scanning Calorimetry*, Dupont Instruments, 1978.
11. F.E. Freeberg, T.G. Alleman, *Anal. Chem.*, 1966, **38**, 1806.
12. I. Mita, I.Imai, H.Kambe, *Thermochim. Acta*, 1971, **2**, 337.
13. H.G. Wiedemann, G. Bayer, *Thermochim. Acta*, 1985, **83**, 153.
14. P.J. Haines, G.A. Skinner, *Thermochim. Acta*, 1982, **59**, 343.
15. H.G. Wiedemann, A.Boller, *Int. Lab.*, 1992, **22**, 14.
16. *Dupont Application Brief*, No. 900B32, 1970.
17. P.J. Haines, *Educ. in Chem.*, 1969, **6**, 171.
18. R.N. Rogers, E.D. Morris, *Anal. Chem.*, 1966, **38**, 410.
19. M.E. Brown, *Introduction To Thermal Analysis*, Chapman & Hall, London, 1988.
20. C.J. Keattch, D. Dollimore, *Introduction to Thermogravimetry* (2nd edn), Heyden, London, 1975.
21. F.W. Wilburn, J.R. Hesford, J.R. Flower, *Anal. Chem.*, 1968, **40**, 777.
22. R. Melling, F.W. Wilburn, R.M. McIntosh, *Anal. Chem.*, 1969, **41**, 1275.
23. F.W. Wilburn, D. Dollimore, J.S. Crighton, *Thermochim. Acta*, 1991, **181**, 173.
24. F.W. Wilburn, D. Dollimore, J.S. Crighton, *Thermochim. Acta*, 1991, **181**, 191.
25. H.J. Borchardt, F. Daniels, *J. Amer. Chem. Soc.*, 1957, **79**, 41.
26. A.W. Coats, J.P. Redfern, *Nature*, 1964, **201**, 68.
27. M.J. Vold, *Anal. Chem.*, 1949, **21**, 683.
28. F.W. Wilburn, R.M. McIntosh, A. Turnock, *Trans. Brit. Ceram. Soc.*, 1974, **73**, 117.
29. S.L. Boersma, *J. Amer. Ceram. Soc.*, 1955, **38**, 281.
30. E.L. Charsley *et al.*, *J. Thermal Anal.*, 1993, **40**, 1415.
31. R.W. Carling, *Thermochim. Acta*, 1983, **60**, 265.
32. A.P. Gray, *Perkin-Elmer Thermal Analysis Study*, #1, Perkin-Elmer Ltd, 1972.
33. J. Pöyhönen, T. Sivonen, M. Hilpela, *Proc. 1st ICTA, Aberdeen*, Macmillan, London, 1965, p. 148.
34. Z.G. Szabo, I.K. Thege, E.E. Zapp, *Proc. 1st ESTA, Salford*, Heyden, London, 1976, p. 272.
35. W. Engel, *Explosivstoffe*, 1973, **21**, 9.
36. D. Chapman, *Chem. Rev.*, 1966, **62**, 433.
37. J.S. Aronhime, *Thermochim. Acta*, 1988, **134**, 1.
38. N.V. Lovegren, M.S. Gray, R.O. Feuge, *J. Amer. Oil Chem. Soc.*, 1976, **53**, 83.
39. E. Kaiserberger, *Thermochim. Acta*, 1989, **151**, 83.
40. J.L. Ford, P. Timmins, *Pharmaceutical Thermal Analysis*, Ellis Horwood, Chichester, 1989.
41. M.J. Hardy, *Proc 7th ICTA, Canada*, Wiley, Chichester, 1982, p. 876.
42. B. Sustar, N. Bukovec, P. Bukovec, *J. Thermal Anal.*, 1993, **40**, 475.
43. G.S. Attard, G. Williams, *Chem. in Brit.*, 1986, **22**, 919.
44. G.W. Gray, *Molecular Structure and Properties of Liquid Crystals*, Academic Press, London, 1962.
45. H.G. Wiedemann, *Mettler Application*, No. 805, 1987.
46. F.D. Ferguson, T.K. Jones, *The Phase Rule*, Butterworth, London, 1966, p. 94.
47. P. Atkins, *Physical Chemistry*, OUP, Oxford, 1978, Ch. 10.
48. E.A. Franceschi, G.A. Costa, *J. Thermal Anal.*, 1988, **34**, 451.
49. P.-Y. Chevalier, *Thermochim. Acta.*, 1989, **155**, 211.
50. A.V. Galanti, R.S. Porter, *J. Phys. Chem.*, 1972, **76**, 3089.
51. H. Jacobson, G. Reier, *J. Pharm. Sci.*, 1969, **58**, 631.
52. D.J.W. Grant, I.K.A. Abougela, G.G. Liversidge, J.M. Padfield, *Anal. Proc.*, 1982, 545 and 549.
53. F. Giordano, G.P. Bettinetti, A. La Manna, A. Marini, V. Berbenni, *J. Thermal Anal.*, 1988, **34**, 531.
54. F. Rodriguez, *Principles of Polymer Systems*, McGraw-Hill, Singapore, 1983.
55. M. Dole, *J. Polym. Sci. (C)*, 1967, **18**, 57.
56. E.M. Barrall *et al.*, *J. Appl. Polym. Sci.*, 1965, **9**, 3061.
57. M.E. Brown, *Introduction to Thermal Analysis* Chapman & Hall, London, 1988, p. 46.
58. M.J. O'Neill, *Anal. Chem.*, 1966, **38**, 1331.
59. B. Wunderlich, M. Dole, *J. Polym. Sci.*, 1957, **24**, 201.
60. B. Wunderlich, *J. Phys. Chem.*, 1965, **69**, 2078.

61. S.Z.D. Cheng, B. Wunderlich, *Thermochim. Acta*, 1988, **134**, 161.
62. A. Everett, *Materials*, Mitchells, London, 1986.
63. Mettler DSC 20 System Reference, Mettler-Toledo.
64. A.R. Ubbelohde, *Quart. Rev.*, 1957, **XI**, 246.
65. B. Wunderlich, M. Bodily, *J. Polym. Sci. (C)*, 1963, **6**, 132.
66. W.P. Brennan, *Perkin-Elmer Thermal Analysis Application Study*, 8 & 11, W.P. Brennan, Perkin-Elmer Ltd, 1973.
67. M.E. Brown, *J. Chem. Ed.*, 1979, **56**, 310.
68. P.W. Atkins, *Physical Chemistry*, OUP, Oxford, 1978, p. 219.
69. G. Widmann, O. Scherrer, *J. Thermal Anal.*, 1991, **37**, 1957.
70. S.T. Glasstone, *Textbook of Physical Chemistry* Macmillan, London, 1951, p. 650.
71. C. Plato, A.R. Glasgow, *Anal. Chem.*, 1969, **41**, 330.
72. P.D. Garn, B. Kawalec, J.J. Houser, T.F. Habash, *Proc. 7th ICTA*, Wiley, Chichester, 1982, p. 899.
73. K.E.J. Barrett, *J. Appl. Polym. Sci.*, 1967, **11**, 1617.
74. J.R. Knox, *Analytical Calorimetry* (ed. R.S. Porter, J.F. Johnson), Plenum, New York, 1968, p. 45.
75. A. Wlochowicz, M. Eder, *Thermochim. Acta*, 1988, **134**, 133.
76. ASTM E698–79: *Arrhenius Kinetic Constants for Thermally Unstable Materials*, ASTM, Philadelphia, 1979.
77. T. Ozawa, *J. Thermal Anal.*, 1970, 2, 301; 1975, 7, 601.
78. M.I. Pope, M.D. Judd, *Differential Thermal Analysis*, Heyden, London, 1977.
79. W. Wendlandt, J.P. Smith, *J. Inorg. Nucl. Chem.*, 1963, **25**, 843, 1267.
80. J.E. House, K. Farran, *J. Inorg. Nucl. Chem.*, 1972, **34**, 1466.
81. G. Beech, C.T. Mortimer, E.G. Tyler, *J. Chem. Soc.*, 1967, 925, 929.
82. R.C. Mackenzie, *Differential Thermal Analysis*, 2 Vols, Academic Press, London, 1970 and 1972.
83. *Recommendations for Testing High Alumina Cement Concrete Samples by Thermo-analytical Techniques*, Thermal Methods Group, RSC, London, 1975
84. H.G. Midgley, *Trans. Brit. Ceram. Soc.*, 1967, **66**, 161
85. J.W. Dodd, K.H. Tonge, *Thermal Methods*, ACOL, Wiley, 1987, Ch. 6.
86. R.C. Mackenzie (ed.), *The Differential Thermal Analysis of Clays*, Mineralogical Soc., London, 1957.
87. D.N. Todor, *Thermal Analysis of Minerals*, Abacus Press, Tunbridge Wells, 1976.
88. C.M. Earnest, *Thermal Analysis of Clays, Minerals and Coal*, Perkin-Elmer, Norwalk, 1984.
89. J.G. Dunn, *Stanton Redcroft Information Sheet*, No. 130, 1978.
90. J.W. Rue, W.R. Ott, *J. Thermal Anal.*, 1974, **6**, 513.
91. J.B. Farmer, A.J.D. Gilbert, P.J. Haines, *Proc. 7th ICTA*, Wiley-Heyden, Chichester, 1982, 650.
92. A.G. Sadler, W.D. Westwood, D.C. Lewis, *J. Can. Ceram. Soc.*, 1971, **21**, 127.
93. S.T. Henderson, P.W. Ranby, *J. Electrochem. Soc.*, 1951, **98**, 479.
94. M.E. Brown, R.A. Rugunanan, *Thermochim. Acta*, 1988, **134**, 413.
95. P.G. Laye, E.L. Charsley, *Thermochim. Acta*, 1987, **120**, 325.
96. T. Ozawa, *Thermochim. Acta*, 1988, **133**, 11.
97. H.-J. Chen, J.H. Sharp, *J. Thermal Anal.*, 1993, **40**, 379.
98. ASTM D 3350–84 and ASTM D 3895–80, ASTM, Philadelphia.
99. J.B. Howard, *Polym. Eng. Sci.*, 1973, **13**, 429.
100. J.M. Barton, *Advances in Polymer Science*, 1985, **72**, 111.
101. R.B. Cassel, *Perkin-Elmer Thermal Analysis Application Study*, 5, Perkin-Elmer, 1973.
102. Setaram File 10: Micro-DSC. Setaram.
103. C.F. Cullis, M.M. Hirschler, *The Combustion of Organic Polymers*, Clarendon Press, Oxford, 1981.
104. G.P. Morie, T.A. Powers, C.A. Glover, *Thermochim. Acta*. 1972, **3**, 259.
105. R.J. Seyler, *Thermochim. Acta*, 1976, **17**, 129.
106. B.Cassel, M.P. DiVito, *Int. Lab.*, 1994, **24**, 19.
107. H.G. Wiedemann, A. Boller, *Int. Lab.*, 1992, **22**, 14.
108. L.C. Thomas, *Int. Lab.*, 1987, **17**, 30.
109. *DuPont Thermal Analysis Application Briefs*, Nos 900B31 and 900B32, 1970.
110. F.R. Wright, G.W. Hicks, *Polym. Eng. Sci.*, 1978, **18**, 378.

111. G.R. Tryson, A.R. Schultz, *J. Polym. Sci., Polym. Phys. Ed.*, 1979, **17**, 2059.
112. T.R. Manley, G. Scurr, *Proc. 2nd ESTA, Aberdeen*, Heyden, London, 1981.
113. K.A. Hodd, N. Menon, *Proc. 2nd ESTA, Aberdeen*, Heyden, London, 1981.
114. P.S. Gill, S.R. Sauerbrunn, M. Reading, *J. Thermal Anal.*, 1993, **40**, 931.
115. M. Reading, D. Elliott, V.L. Hill, *J. Thermal Anal.*, 1993, **40**, 949.
116. A.F. Barnes, M.J. Hardy, T.J. Lever, *J. Thermal Anal.*, 1993, **40**, 499.
117. *TA Instruments Ltd Application Brief*, MDSC-1, 1993.

Bibliography   The general reference texts given in Chapter 1 all contain substantial sections on DTA and DSC. In addition, the following are more specific to these techniques:

R.C. Mackenzie, *Differential Thermal Analysis* (2 Vols), Academic Press, London, 1970 and 1972.
W.J. Smothers, J. Chiang, *Differential Thermal Analysis*, Chemical Publishing Co, New York, 1958.
M.I. Pope, M.D. Judd, *Differential Thermal Analysis*, Heyden, London, 1977.
J.L. McNaughton, C.T. Mortimer, *Differential Scanning Calorimetry*, Perkin-Elmer, 1975, Butterworth (Vol. 10 of IRS Physical Chemistry Series 2).
E. Kaisersberger, H. Möhler, *DSC on Polymeric Materials*, Netzsch Annual for Science and Industry, Vol. 1, 1991,
E. Kaisersberger, S. Knappe, H. Möhler, *TA for Polymer Engineering: DSC TG DMA*, Netzsch Annual, Vol. 2, 1993, Netzsch, Selb.

Thermomechanical, dynamic mechanical and associated methods

4

M. Reading and P.J. Haines

The mechanical properties of materials are an essential guide to their suitability for particular usage, and can indicate how the material has been treated before testing. The molecular nature of the material will be most important in determining the mechanical properties. For example, the behaviour of plastics will be very greatly influenced by their chemical structure, their blending and the way in which they have been fabricated.

The mechanical methods divide into two classes, depending on whether the forces applied are constant or varied. Very often, the parameters and properties which are measured are specific to the method, but can still be extremely useful in comparing materials.

4.2.1 *Thermomechanical analysis (TMA)*

This is a technique in which the *deformation* of the sample under non-oscillating stress is monitored against time or temperature, while the temperature of the sample, in a specified atmosphere, is programmed.

The stress may be compression, tension, flexure or torsion, and if the stress is too low to cause deformation, TMA monitors the *dimensions* of the sample and this role can be called *thermodilatometry*.

If the stress is oscillating, the technique is called *dynamic-load thermomechanical analysis* (DLTMA) [1].

4.2.2 *Dynamic mechanical analysis (DMA)*

This is a technique in which the storage modulus and loss modulus of the sample, under oscillating load, are monitored against time, temperature or frequency of oscillation while the temperature of the sample in a specified atmosphere is programmed. The experiment is normally carried out in such a way as to maintain the dynamic strain constant, although constant stress experiments are also used sometimes. When constant strain is used, the deformation is usually of the order of a few percent or less.

Certain specific methods, such as *torsional braid analysis* and *torsional pendulum studies* may be incorporated into this definition, and *dielectric thermal analysis* (DETA) gives results that may complement those of DMA.

4.2.3   *Mechanical moduli*

If any sample is subjected to a force, it may behave in a variety of ways [2,3]. A large force, suddenly applied, will often break the material, but a small force will deform it. Liquids will flow when a force is applied, depending on their *viscosity*, η. Some solids may deform *elastically*, returning exactly to their former shape and size when the force is withdrawn. Others may behave *viscoelastically*, showing behaviour which incorporates both flow and elastic deformation. With many materials, there is an *elastic limit* above which the material undergoes plastic deformation which is *irreversible*. Further increase in load eventually causes fracture.

The parameters, symbols and terminology that will be used for studying mechanical properties must be established [2,3,4]. The symbols are illustrated in Figure 4.1.

The *stress* is the force applied per unit area. This may be :

	a normal tensile stress:	σ	$= F/A$
or	a tangential, shearing stress:	τ	$= F/A$
or	a pressure change:	Δp	$= F/A$

All of these will have units of N/m^2, or Pa.

This stress will cause a deformation measured by the *strain*, which is the deformation per unit dimension, for example:

	tensile strain or elongation:	ϵ	$= \Delta l/l$
or	shear strain:	γ	$= \Delta x/y$
or	volume or bulk strain:	θ	$= \Delta V/V$

Strain has no units.

For an *elastic* material, Hooke's law applies and strain is proportional to stress, the constant being the *modulus*:

Modulus = stress/strain

Tensile, or Young's modulus:	E	$= \sigma/\epsilon$
Shear modulus:	G	$= \tau/\gamma$
Bulk (or compression) modulus:	K	$= \Delta p/(\Delta V/V)$

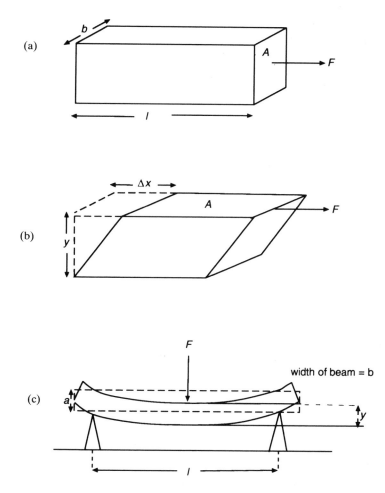

Figure 4.1 Schematic of types of deformation: (a) tensile; (b) shear; (c) three-point bending.

The *distortion*, where increasing the length l of a material also decreases the breadth b, is measured by Poisson's ratio:

Poisson's ratio: $\quad v = (\Delta b/b)/(\Delta l/l)$

This is generally between 0 and 0.5.

The moduli are related to each other for isotropic materials:

$$G = E/(2(1 + v)).$$
$$K = \tfrac{1}{3}(EG/(3G + E)) = E/(3(1-2v))$$

Penetration occurs when the applied stress causes the probe to sink into the material, and a penetration modulus related to the depth of

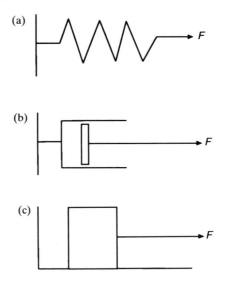

Figure 4.2   Model components for viscoelastic behaviour: (a) spring or Hookean body; (b) dashpot or Newtonian body; (c) sliding weight or Saint-Venant body.

penetration, the probe shape, the sample shape and the force may be derived.

Three-point bending is shown in Figure. 4.1(c). If a bar of width b and depth a is supported by two knife edges a distance l apart, and loaded centrally with a force F, it may be shown that the deflection y is given by:

$$y = Fl^3/(4ba^3E)$$

Viscous flow occurs when the shear stress τ causes adjacent layers to move relative to each other. The velocity gradient across the layers is given by the rate of change of the shear strain, $d\gamma/dt$. Ideal liquids show a proportionality between these by Newton's law:

$$\tau = \eta \, (d\gamma/dt)$$

There are several types of material which show 'non-Newtonian' behaviour, and do not obey the above law. For example, paints may show an apparent decrease in η as the shear stress is increased.

The more complex behaviour of materials can be represented by combining the models above. We may represent the elastic part by a spring (or 'Hookean body'), the viscous part by a dashpot (or 'Newtonian body') and add another element provided by the sliding weight (or 'Saint-Venant body') which will move when the frictional forces are overcome. These are shown in Figure 4.2.

Combining the dashpot and spring in series gives a 'Maxwell body' which

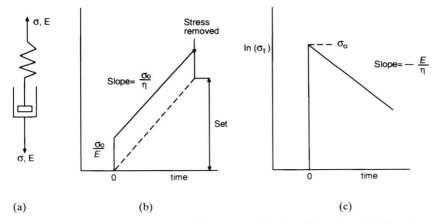

Figure 4.3 Viscoelastic behaviour: (a) the Maxwell body with spring and dashpot in series; (b) creep behaviour; (c) stress relaxation.

will allow adequate quantitative explanation of viscoelastic phenomena. Since the stress, σ, is the same and the elongations, ϵ, are additive,

$$d\epsilon/dt = d\epsilon_1/dt + d\epsilon_2/dt = (1/E)\,(d\sigma/dt) + \sigma/\eta$$

Creep is the gradual, irreversible elongation of a sample when subjected to a stress due to a combination of elastic behaviour and viscous flow. The application of the initial stress will cause an initial elongation, ϵ_0, which is then increased by viscous flow. When the stress is removed, only the deformation of the spring is recovered, and the dashpot flow gives a permanent 'set', as shown in Figure 4.3(b):

$$\epsilon(t) = \epsilon_0 + \sigma_0 t/\eta$$

Stress relaxation may be analysed by considering a fixed elongation at the start, and integrating the equation for the Maxwell body to see how the stress changes with time. We find that:

$$\ln(\sigma_t/\sigma_0) = -Et/\eta$$

This behaviour is illustrated in Figure 4.3(c)

In the case of an ideal elastic material, the deformations are exactly reversible. If there is any viscoelasticity, however, the moduli become *complex* and contain two parts. For example, in the case of the tensile modulus, one is the *storage modulus* E', and the other the *loss modulus* E'':

$$E^* = E' + iE''$$

where $i = \sqrt{-1}$.

The ratio of these two moduli gives the loss tangent,

$$\tan(\delta) = E''/E'$$

The thermal effects on materials may change all these parameters, and also cause the material to expand. The coefficient of thermal expansion is

$$\alpha = (\mathrm{d}l//l)/\mathrm{d}T$$

This may also change with temperature, or with the nature of the material, especially at such events as the glass transition.

**4.3
Thermomechanical
analysis**

4.3.1 Apparatus

While some equipment uses optical or mechanical measurements, many modern TMA systems use a linear variable differential transformer (LVDT) to produce an electrical signal from a linear movement [5]. This consists of two mutual inductances with a common ferromagnetic core. The magnetisation of the core by the primary coil induces a current in the secondary coils. If a pair of symmetrically placed secondary coils connected in series opposition are used, then equal but opposite signals are produced when the core is at the centre point. When the core is displaced the net signal increases. By using a phase-sensitive detector, a good linear relationship between displacement and output is obtained.

The total equipment, shown in Figure 4.4, has the typical arrangement

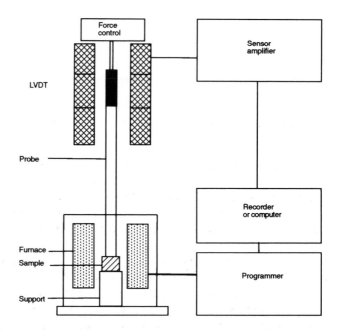

Figure 4.4   Schematic of a TMA system.

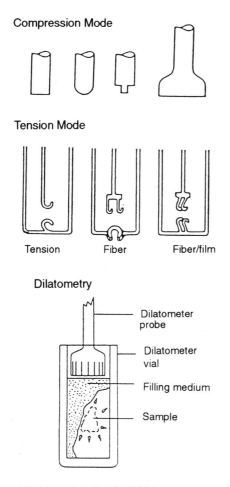

Figure 4.5   Typical probes for TMA measurement [6].

of all thermoanalytical equipment. The LVDT is attached to a quartz probe which rests on the sample, enclosed in a controlled atmosphere, and a sample holder if necessary, and the whole assembly is enclosed in the programmer-controlled furnace. For work below room temperature, the assembly is cooled by placing a Dewar flask of liquid nitrogen around it.

The force exerted on the sample by the probe will affect the thermo-analytical results obtained. For that reason an additional unit is added which can control the force, using mechanical or computer systems, so that we may apply a compressive force, or tension, or a zero force for dilatometry, or an oscillatory force for dynamic measurements (Figure 4.5).

PROBES

Probes of higher area will exert less *pressure* (since $p=F/A$) on the sample, while those with a sharply pointed, small area or a rounded end will tend to penetrate the sample. Special attachments will allow the same apparatus to be used for flexural or bending measurements and with films. The probes may be loaded manually or electromagnetically using the force control system.

Dynamic or oscillating loads may be used and will often give further information of thermal changes (DLTMA).

Viscosity measurements may be made using a *parallel plate rheometer* (PPR) arrangement. This may be either an oscillating plate method, or a method where the viscous behaviour is studied by measuring the approach of two parallel plates moving towards each other with a viscoelastic medium between them [7].

CALIBRATION

The calibration of the probe (Figure 4.6) for its response to length and changes in length is generally carried out with standard length pieces of metal or ceramic which have been measured independently.

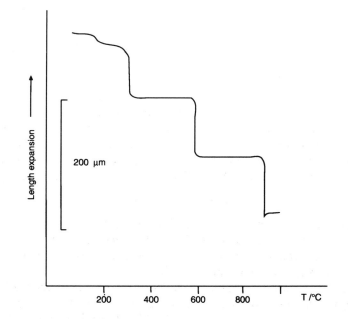

Figure 4.6 Calibration graph for TMA. Steps correspond to the melting of tin (232 °C), aluminium (660 °C) and silver (960 °C).

Table 4.1

Material	Coefficient of linear expansion ($\alpha \times 10^6$/K)
Fused silica	0.4
Pyrex glass	3
Aluminium	23
Epoxy resins	45–65
Polypropylene	60–100

Calibration for temperature is complicated by the distance between the sample and the measuring thermocouple in some larger apparati. Representative samples may be 5 mm in thickness. A check may be carried out using discs of ICTAC standards, such as indium, tin and lead, separated by inert discs of alumina. When this 'sandwich' is placed on the holder and the probe placed on top, the melting of each standard as the apparatus is heated is shown by the penetration of the probe.

Recent work [8] and round-robin testing of the calibration with metals and of measurements with standard polymers has shown reasonable agreement, although Willcocks [9] has pointed out that the probe force ought to be calibrated also.

4.3.2 *Applications*

COEFFICIENTS OF EXPANSION

Strictly, this is 'thermodilatometry', since the probe force used should be zero. Since the apparatus and techniques are similar to the other applications, both will be considered together. The probes and supports are generally of fused silica whose coefficient of expansion is very low in comparison with most other materials (Table 4.1).

For a single material, undergoing no transitions, the average coefficient of expansion between two temperatures may be measured, as shown in Figure 4.7. An 'instantaneous' value of α may be found from the slope of the TMA trace, also called the 'DTMA'. Here we may use the equation:

$$dl/dt = l_0 \cdot \alpha (dT/dt)$$

For ceramics, the expansion of the glaze must be matched to that of the body and this has been studied by dilatometry. The coefficient of expansion of ceramics is also dependent on the quartz content, the firing temperature and the quartz grain size [10].

Laminated electronic printed circuit boards are often checked for their dimensional stability by TMA and the coefficient of linear expansion of electrical insulating materials determined by TMA is a standard method ASTM D3386–84 [11].

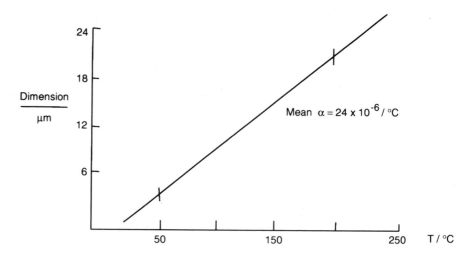

Figure 4.7   Coefficient of expansion of aluminium between 50 and 200 °C (5 mm sample, 10 K/min, static nitrogen, 0.01 N load).

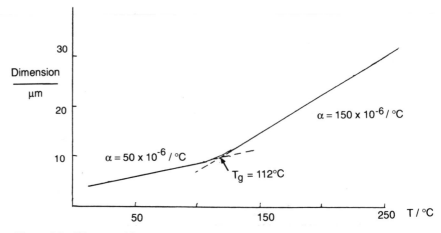

Figure 4.8   Glass transition temperature by TMA. Sample: printed circuit board, 1 mm, 10 K/min, static nitrogen, 0.01 N load.

GLASS TRANSITION TEMPERATURE

At the T_g (Figure 4.8) the heat capacity and coefficient of expansion change abruptly. Under small load α may be measured below and above the T_g and the lines extrapolated back to give a single value. The heating and cooling rates will affect this temperature, as well as the molecular weight, degree of cure, plasticiser content and orientation.

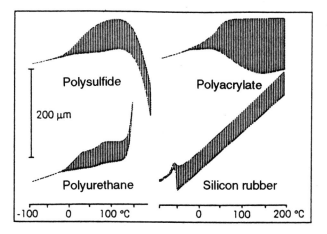

Figure 4.9 Typical DLTMA curves of four elastomers (sealing compounds) in the temperature range −90 to 200 °C. Sample thickness 2–3 mm, 10 K/min alternating load of 0.00 and 0.04 N.

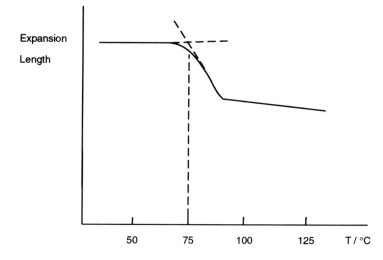

Figure 4.10 Glass transition and softening in penetration mode. Sample of coated material, 10 K/min, 1 N load.

The oscillating load technique (DLTMA) is somewhat better for detecting small changes, for example in plasticised elastomers (Figure 4.9) [12].

If a higher force is used, particularly with the *penetration* probe a trace similar to Figure 4.10 may be obtained. Below the T_g little penetration is

Table 4.2

	VICAT/°C		DTUL/°C	
Sample	ASTM	TMA	ASTM	TMA
Unmodified PVC	84	86	75	73
Modified PVC	103	106	91	90
Impact modified PVC	85	84	73	73

shown, but above this temperature the polymer softens and the probe sinks into the material. At the melting point, complete penetration occurs.

This experiment is particularly useful when the sample is coated onto a substrate, for example an epoxy coating on a container; the T_g must be selected by controlling the cure so that the coating will not crack in use nor will the coating flow [13]. Similar experiments are used for a resin coating on a wire.

SOFTENING TEMPERATURES

Standard ASTM methods are available for detecting the softening of a polymer. The VICAT test involves the measurement of the temperature at which a particular penetration is obtained for a specified load on a particular sample. The deflection temperature under load (DTUL) is similar, involving the bending of a sample supported at its ends (Table 4.2). It has been shown by Yanai *et al.* [14] that by choosing suitable loads and conditions these indices may be matched by TMA measurements.

The softening of other materials such as soaps and chocolate may also be studied using TMA with the penetration probe.

SOLVENT SWELLING OF POLYMERS (FIGURE 4.11)

The sample and probe are immersed in a suitable solvent, for example, a polystyrene–polybutadiene copolymer may be immersed in toluene. The gradual uptake of the solvent causes the sample to swell, but also contaminates the solvent, which may have to be renewed. The swelling may be used to calculate the cross-link density using the Flory–Rehner equation [15]:

$$n = -[\ln(1-v_2) + v_2 + \varkappa_1 \cdot v_2^2 / v_1 [v_2^{1/3} - v_2/2]$$

where n is the cross-link density (mol/cm³), v_2 is the volume fraction of polymer at equilibrium, v_1 is the molar volume of solvent and $\varkappa_1$ is the polymer solvent interaction parameter.

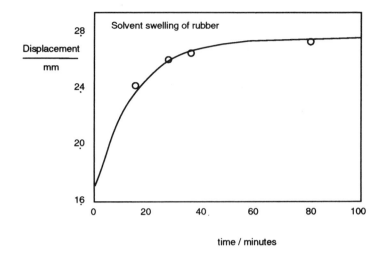

Figure 4.11   Swelling of polyisoprene, immersed in toluene. Isothermal at 40 °C in static nitrogen, 0.1 N load.

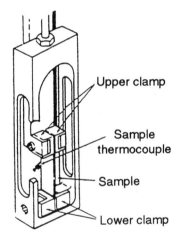

Figure 4.12   Film and fibre attachment for TMA [6].

The expansion, penetration and creep of the samples may all be studied by TMA.

FILMS AND FIBRES

Film samples may be attached by small clamps or clips, and fibre samples by hooks to a special attachment similar to that shown in Figure 4.12.

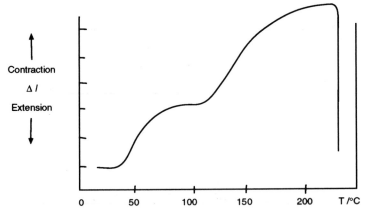

Figure 4.13 Nylon fibre in extension mode, 10 K/min [16].

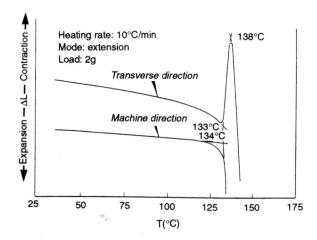

Figure 4.14 Blown polyethylene film. Film and fibre probe, 10 K/min [16].

The sample is generally held under slight tension and the movement of the probe will indicate stretching or shrinkage of the film or fibre. This is especially useful to show transitions near the T_g. The shrinkage of fibres or films often starts just above the glass transition temperature, as the molecular arrangement is more free to alter. Figure 4.13 shows a nylon fibre which shows slight expansion to T_g at 20 °C, then shrinkage to about 100 °C when there is a plateau, possibly due to moisture loss. Further shrinkage continues until melt flow occurs near 250 °C.

Work on blown polyethylene film (Figure 4.14) [16] has shown the effects of orientation, since there are distinct differences between samples

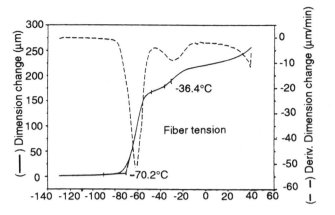

Figure 4.15   Engine oil sample coated onto cotton fibre. The TMA curve (full line) shows a glass transition at about $-70\ °C$ and wax dissolution around $-40\ °C$. The derivative curve (dashed line) makes these events clearer. This experiment is run in tension mode, which gives better sensitivity than the penetration mode [17].

cut from the 'machine direction' in which the film is travelling and the 'transverse direction' perpendicular to it.

One method of measuring transitions in liquids is to add them onto a cotton fibre in the attachment (Figure 4.15) and study their combined behaviour, rather as in torsional braid analysis (see p. 149).

PHASE TRANSITIONS

The molecular structure of different polymorphic forms means that they have different densities, and different coefficients of expansion. This may be detected using the normal expansion probe or a special 'dilatometric attachment' where the expansion of the sample is transmitted to the probe through an inert packing. Figure. 4.16 shows a trace for sulphur using a normal probe. The expansion above 95 °C is due to the α–β phase change, and the melting causes a contraction as the sample flows.

The phase transitions of sodium and potassium nitrates and of potassium perchlorate as well as the polymorphic crystal transitions of organics give TMA traces with a 'step' at the transition temperature.

The α–β quartz transition at 573 °C, which gives a clear peak on the DTMA trace, was shown to be a suitable method for determining the quartz content in ceramics [10].

SINTERING

The compaction and sintering of high-temperature materials may be studied by TMA. Figure 4.17 shows the sintering of a refractory clay block

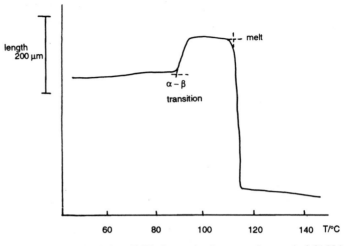

Figure 4.16   TMA of sulphur (5 K/min, static air, expansion mode 0.01 N load).

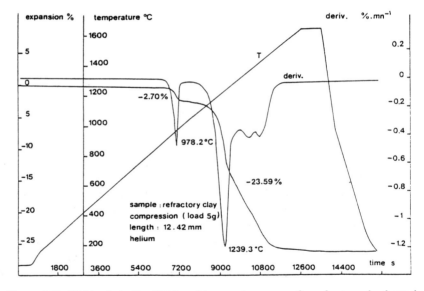

Figure 4.17   TMA, derivative TMA and temperature curves for refractory clay heated under conditions to reproduce industrial furnaces [18]. (Courtesy SETARAM).

under conditions designed to reproduce industrial furnaces. Very little change occurs up to 900 °C but a small shrinkage of 2.7% takes place at about 980 °C, and a much larger one of 23.6% between 1100 and 1600 °C. The derivative curve shows this most clearly.

A 'quasi-isothermal' approach has been used to study sintering of UO_2

powders [19]. The overall heating rate is controlled by the shrinkage rate. When this rate becomes larger than a preset limit, the heating is stopped and the shrinkage continues isothermally until the rate falls below the limit again.

4.3.3 *Chemical reactions*

INORGANIC HYDRATES

Using a dilatometer adapted to record the electrical conductance of the sample simultaneously with the dimensional change, Karmazsin *et al.* [20] showed that the water losses from hydrated copper sulphate gave corresponding shrinkages and increases in conductance.

An investigation of the uses of TMA as well as DSC and TG in the selection of binding and construction materials [21] showed the distinction between forms of calcium sulphate hemihydrate, and the effects of the constituents of cement mortar (Portland cement, sand and water) on the TMA curve.

POLYMER CURE

There are several ways of using TMA to study polymer cure. If the effect of cure on the glass transition temperature has already been established, TMA could be used to measure T_g after various cure times. As a polymer cross-links, at a certain conversion it forms a 'gel' or network structure and the viscosity rises considerably. The 'gel time' of a resin may be found using the dynamic load technique (DLTMA). With the resin contained in a suitable container, held isothermally, the probe force is changed between about -0.05 and $+0.05$ N. Before the resin gels, the oscillations are very large. When the sample gels, its viscosity suddenly increases and the probe can no longer oscillate freely, so the trace becomes level [22].

Using the parallel plate rheometer (PPR) attachment, the gel time of a prepreg material has been investigated. When heated at 10 K/min, the material first expands slightly, then softens at about 70 °C and eventually, after around 8 min at about 85 °C, gelation occurs and the probe stops moving, as shown in Figure 4.18 [23].

4.4 Dynamic mechanical analysis

If an oscillatory sine-wave stress is applied to a perfectly elastic solid, the deformation and the strain will be exactly in phase with the stress. When the same oscillatory stress is applied to a viscoelastic solid, the strain lags behind the stress and is out-of-phase by an angle δ. The complex dynamic modulus (E^* in extension mode) must be used:

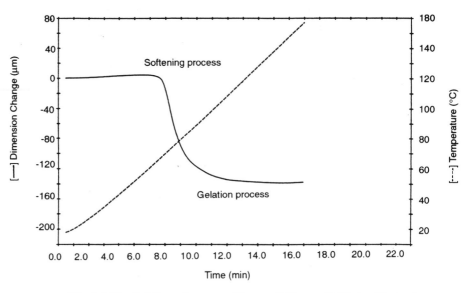

Figure 4.18   Gel time of prepreg. 5 layers, 0.54 mm, 10 K/min, 10 g load.

$$E^* = E' + iE''$$

where E' is the storage modulus, E'' is the loss modulus and $i = \sqrt{-1}$.

The loss tangent is the tangent of the phase angle and

$$\tan \delta = E''/E'$$

These are illustrated in Figure 4.19.

The values of the moduli will change with temperature as the molecular motions within the material change. Chain and side-chain motions of polymers and especially glass transitions will affect the moduli and tan δ and the increase of frequency will also show a shift of the moduli and tan δ up the temperature scale. These effects may be shown on a 3-dimensional diagram, as in Figure 4.20.

The design of instrument may allow *fixed frequency* operation where the frequency and amplitude are selected by the operator, or *resonant frequency* operation, where the instrument allows the sample to oscillate at its natural resonant frequency. Instruments like the torsional pendulum and torsional braid analyser generally operate in this mode.

Several different types of dynamic mechanical analysers (or dynamic mechanical thermal analysers) are commercially available, and their designs vary considerably. Ideally, it should be possible to make measurements with several different vibrational frequencies, with different clamping arrangements, at different temperatures, and to apply the technique to materials with a very wide range of moduli. The frequency range used is generally between 0.01 and 100 Hz.

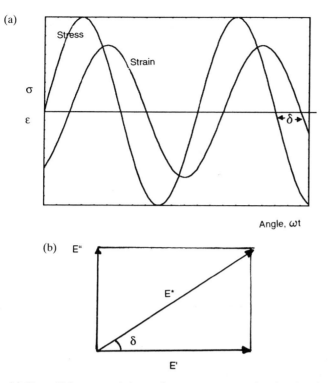

Figure 4.19 (a) Sinusoidal stress and the strain response curve, showing the phase angle lag, due to the viscoelastic behaviour. (b) The relationship between the complex modulus E^* and its real component E' and imaginary component E''.
E' = storage modulus, E'' = loss modulus.

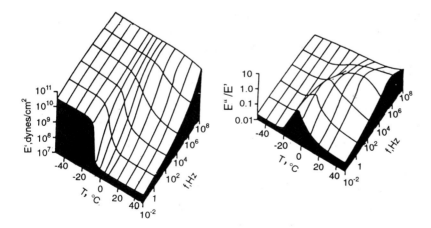

Figure 4.20 Effect of temperature and frequency on the moduli of a styrene–butadiene rubber over the main T_g region. Note: 10 dyne/cm^2 = 1N/m^2 [24].

4.4.1 Apparatus

One approach is to use a design similar to a TMA and to achieve the necessary constant dynamic strain through computer control of the linear induction motor. A second approach is to mount the mechanism responsible for generating the cyclic deformation on a carriage that is moved in order to accommodate changes in the overall sample size.

The furnace arrangement surrounds the mechanism and sample, and cooling is achieved by pumping liquid nitrogen through the chamber surrounding the furnace.

DMA CONFIGURATIONS

Modern DMA systems can operate in many modes, as illustrated in Figure 4.21(a) and (b).

Flexure may be measured by single and double cantilever bending and three-point bending. Shear and extension modes may also be studied by simply changing the clamping configuration. Torsion requires an instrument dedicated to this mode. Flexure is preferred for high-modulus materials, as measurable deformations may be achieved with comparatively modest forces.

It is advisable to avoid the use of very high applied forces as these may cause the clamping system to deform significantly and thus introduce errors.

Extension is clearly preferable for fibres and thin films. To avoid buckling, these sample types should be kept under a constant static stress so that at no point in the deformation cycle does the instrument attempt to compress the sample. If a constant dynamic stress experiment is being carried out, the dynamic stress will change as the modulus of the sample changes. The static stress must then change with the dynamic stress to avoid buckling. Usually they are kept at a fixed ratio of 1.1–2 so that they remain in concert.

CALIBRATION

There are a large number of unresolved problems with regard to accurate calibration of DMA apparatus for temperature, storage and loss moduli. The sample may be large, which presents problems with thermal transfer and temperature monitoring. There are no universally accepted reference materials for DMA calibration. Pure metals such as indium may be used for temperature calibration but they have very different thermal conductivities from polymer samples. It has been reported that it is rarely possible to achieve agreement between instruments of better than 5 K

(a)

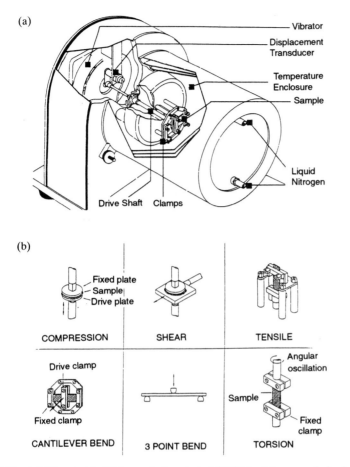

Vibrator

Displacement
Transducer

Temperature
Enclosure

Sample

Liquid
Nitrogen

Drive Shaft   Clamps

(b)

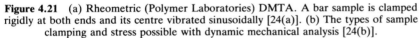

Fixed plate
Sample
Drive plate

COMPRESSION   SHEAR   TENSILE

Drive clamp   Angular
                                         oscillation

Fixed clamp   Sample

CANTILEVER BEND   3 POINT BEND   Fixed
                                        clamp

                                        TORSION

Figure 4.21 (a) Rheometric (Polymer Laboratories) DMTA. A bar sample is clamped rigidly at both ends and its centre vibrated sinusoidally [24(a)]. (b) The types of sample clamping and stress possible with dynamic mechanical analysis [24(b)].

and when using the liquid nitrogen cooling accessory, and the same reference standard (PMMA), differences of 10–15 K were sometimes obtained [9]. The clamping of the material may also affect the properties measured.

Currently, the best practice is to follow the manufacturer's calibration procedures regularly to ensure calibration of the temperature sensor with reference metals. This should then be checked using a suitable stable reference polymer to detect calibration drift.

This should ensure that, while absolute values will have to be treated with some caution, particularly with high-modulus materials, trends in results from the same instrument under the same conditions can be interpreted with confidence.

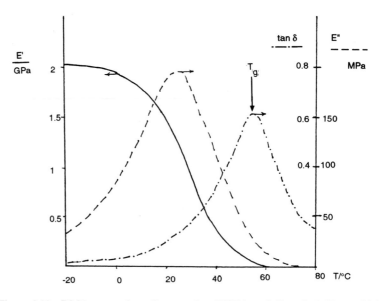

Figure 4.22   DMA curves for a fire-retardant PVC board. Sample 1.67 mm thick, 5 K/min, nitrogen, 10 Hz.

4.4.2 Applications

GLASS TRANSITION TEMPERATURES

As a polymer passes through its glass transition temperature, T_g, the storage modulus usually decreases by two or three orders of magnitude, and the tan δ goes through a maximum. Why should this be so?

The decrease in the moduli occurs when there is main chain molecular motion and the maximum in tan δ occurs when the frequency of the forced vibration coincides with the frequency of the diffusional motion of the main chain. An example of the DMA of a glass transition is shown in Figure 4.22. DMA is very much more sensitive than DSC or other thermal techniques for studying the glass transition.

BETA AND OTHER TRANSITIONS

For many polymers, in addition to the glass transition, there are also secondary transitions observed at lower temperatures. By convention, going down in temperature from the melt, the glass transition is also referred to as the alpha transition, then at lower temperatures, the beta, then gamma. This is illustrated in Figure 4.23.

For example, in amorphous poly(vinyl chloride) the alpha or glass

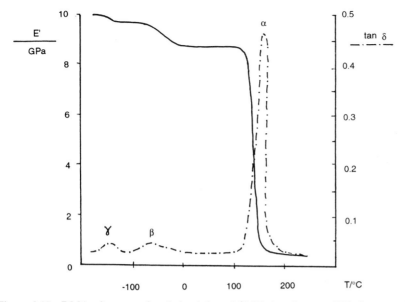

Figure 4.23   DMA of epoxy printed circuit board (5 K/min, nitrogen, 7 Hz frequency).

transition is associated with relaxation of the main backbone chain and occurs at about 88 °C at 1Hz, while the beta transition gives a broad tan δ peak at about −35 °C and is due to motion of small groups, and the gamma peak at −110 °C is also broad and due to motion of small chain segments.

The effects of moisture on these transitions may be investigated by DMA. For example, for nylon 66, a polymer with polar groupings, the glass transition is shifted from 70 °C under dry conditions to about 0 °C under 100% humidity.

The beta and gamma transitions are affected to a smaller extent.

RELAXATION KINETICS

DMA curves are affected by the frequency of the applied deformation, as shown in Figure 4.24. These curves can be analysed using an Arrhenius-type expression, taking the temperature T_{td} at which the tan δ reaches its maximum over a range of frequencies:

$$\ln(f) = \ln(A) - E/RT_{td}$$

where f is the frequency, A is a constant and E is the activation energy for the molecular motion.

An Arrhenius plot of $\ln(f)$ against $(1/T_{td})$ can thus yield the activation energy for what is termed the 'relaxation process'. For beta and lower

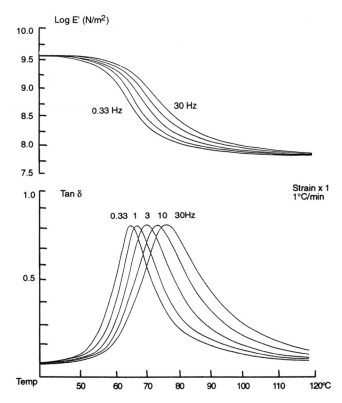

Figure 4.24 DMA curves for a storage modulus E' and tan δ for a filled polymer. The results for all the frequencies were obtained by multiplexing during a single scan at 1 K/min.

temperature transitions this expression often provides a good model. Near the T_g, however, the expression does not apply. Although over a small interval it may appear to give a linear plot, over a broad range it can be seen to be curved [25].

A more appropriate formula for the alpha process is the Williams–Landel–Ferry (WLF) equation [26]. This requires the selection of a reference temperature T_0:

$$\log(a) = - C_1 \cdot (T - T_0) / (C_2 + (T - T_0))$$

where C_1 and C_2 are constants.

The data at other temperatures, T, are then shifted along the $\ln(f)$ axis by a shift factor, $\ln(a)$, until a 'master curve' is produced (see Figures 4.25(a) and (b).)

In complex system including semicrystalline polymers, whether a particular transition is a glass transition or a secondary transition can be investigated by determining whether it exhibits Arrhenius or WLF behaviour.

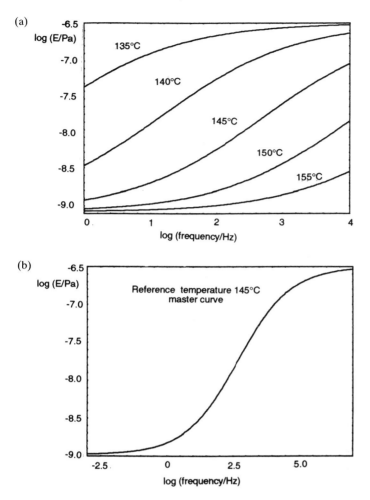

Figure 4.25 (a) A series of curves for E' as a function of frequency derived from multiplexed DMA curves over a range of temperatures. (b) A master curve derived from the data of Figure 4.25(a) using the WLF equation to shift the curves along the log (frequency) axis to correspond with the 145 °C reference curve.

POLYMER MISCIBILITY

DMA is often used to examine the degree of phase separation to be found when two polymers are mixed together. Where two distinct T_g transitions are found that have the same values as the pure materials, then complete phase separation can be inferred. If a single glass transition peak is observed at a temperature intermediate between the values for the two pure materials, then the polymers are generally said to be miscible and intimate mixing has occurred. A common expression that is used to relate

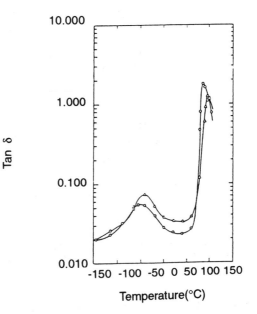

Figure 4.26 DMA curves showing the effect of adding different levels of rubber to a cross-linked matrix [31].

the T_g of the blend to the composition of the blend and the values for the components is the Fox equation [27]:

$$1/T_{g,C} = W_A/T_{g,A} + W_B/T_{g,B}$$

where $T_{g,C}$, $T_{g,A}$ and $T_{g,B}$ are the glass transition temperatures of the blend, component A, and component B, respectively, W_A is the weight fraction of A and W_B is the weight fraction of B.

RUBBER TOUGHENING

The inclusion of low T_g domains within a brittle matrix of higher T_g is a well-established method of obtaining tough materials. Figure 4.26 illustrates how the amount of the rubber component influences the height of the lower temperature transition for a cross-linked epoxy resin modified with a carboxyl-terminated acrylonitrile–butadiene rubber. The strength of the lower temperature transition can provide an indication of the fracture resistance of the material [28].

CHARACTERISING CROSS-LINKING

An important feature of cross-linking systems is that of gelation, the point at which a true network is first formed and causes the viscosity to rise

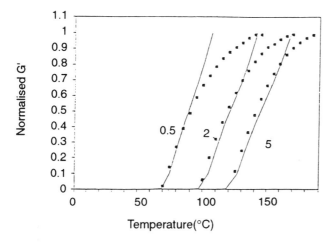

Figure 4.27   DMA results showing the cure of a cross-linking system at different heating rates (squares) with results from a computer model (lines) [31].

towards infinity. This is then followed by a build-up in cross-link density and consequently an increase in the storage modulus.

Prepregs are systems that comprise a reinforcing material, such as carbon fibres, impregnated with a cross-linkable polymer. This type of material can be formed and then heated to cure the system to make a light, rigid product with excellent mechanical properties. DMA is well adapted to subjecting a prepreg sample to a heating schedule, similar to that encountered by the real system, and to following the cure performance.

Gelation is indicated not only by a rapid rise in the storage modulus, E', but also by a peak in tan δ. As the cross-link density increases, the glass transition temperature rises and may reach the cure temperature. When this happens the sample vitrifies. This is also marked by an increase in storage modulus and a peak in tan δ. In this way the transformation of the prepreg from a pliable material to a rigid composite can be characterised.

DMA can be used to monitor the cross-linking of liquids by impregnating them onto an inert support such as a glass fibre braid. In this way, they become similar to the prepregs described above and their cross-linking can be followed in a similar way. The early work in this area was carried out using a *torsional braid analyser* (TBA) developed by Gilham [29], but modern instruments usually use a shear or flexural deformation. With this kind of experiment, it is not generally possible to measure the absolute modulus of the polymer as it is difficult to relate the modulus of the polymer–glass fibre composite to that of the polymer by itself. Figure 4.27 shows some recent results from a study of the curing of an isocyanate–hydroxy system at various heating rates. It can be seen that the DMA results show the gel point clearly and the subsequent build-up in modulus, and, therefore, cross-link density. From this type of experiment kinetic parameters such as the activation energy may be calculated [30].

When a material is sufficiently cross-linked to form a solid with a reasonable degree of mechanical integrity, then its modulus above the glass transition temperature can be used to measure its cross-link density using the formula:

$$G = \rho RT/M_c$$

where G is the equilibrium shear storage modulus, ρ is the density, R is the molar gas constant, T is the temperature (K) and M_c is the average molecular weight between cross-links. The dynamic storage modulus at low frequencies will approximate to the equilibrium value (strictly the zero frequency storage modulus). In this way, the mechanical properties measured by DMA can be related to the polymer structure.

STUDYING 'PROBLEM SAMPLES'

Supported techniques can also be valuable for the investigation of materials that pose difficulties when presented as an unsupported solid, either because they creep too much to be securely clamped, or because they are too brittle. Another advantage is that phase-separated materials in which the matrix has the lower T_g can be investigated. In an unsupported DMA experiment, as the matrix goes through its T_g the sample becomes very soft and may simply flow out of the clamping system before the response of the higher T_g occluded phase can be probed.

Using a supported technique this problem can be overcome. In general, with supported techniques, however, care must be taken that there are no very strong interactions between the support material and the polymer, as these could distort the results, for example, by increasing the apparent glass transition temperature of any strongly interacting component, or causing it to be enriched at the polymer–support interface.

CHARACTERISING FILM FORMATION

Another type of supported technique is particularly useful for looking at film formation based on coating a thin metal shim. A coating is spread on the surface of the shim which is then placed in the DMA in a three-point bending configuration [31].

Figure 4.28 shows the results of an experiment carried out on an emulsion paint. The drying behaviour and solvent loss are followed by close packing and coalescence of the latex particles. This can be followed clearly on the storage modulus curve. The probe position indicates the

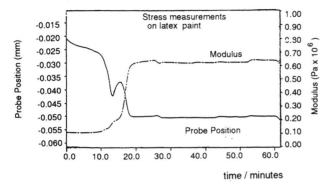

Figure 4.28 DMA results showing the drying behaviour of a latex paint, including the build-up of internal stress [31].

degree of curvature of the coated shim, and therefore serves as a measure of the internal stress building up within the coating.

This information is useful for studying the cracking behaviour or adhesion loss in coating systems. The advantage over the braid technique, in addition to the stress data obtained, is that it more nearly approximates to the real world condition for coatings since the diffusion of solvents to the surface may be impeded from a braid a few millimetres thick when compared to a coating applied to a surface. The effects that might arise due to interactions with the support are also minimised.

4.5 Dielectric thermal analysis

This method, known as DETA, is a complementary method to DMA, where a sinusoidal oscillating *electrical* field is applied to the sample. The sample will behave as a dielectric, and instead of the complex mechanical moduli, DETA measures the *complex dielectric permittivity*, ϵ^*. The permittivity is the constant in the equation for electrical capacitance. For example, with a parallel plate capacitor in which plates of area A are separated by a distance d by a dielectric material, the capacitance C is given by:

$$C = \epsilon A/d = \epsilon_r \cdot \epsilon_0 \, (A/d)$$

where ϵ is the permittivity of the dielectric material, ϵ_0 the permittivity of free space, and ϵ_r ($= \epsilon/\epsilon_0$) is the relative permittivity (dielectric constant). When the permittivity is measured with an a.c. field, it varies with the frequency of that field. The permittivity is now complex:

$$\epsilon^* = \epsilon' - i\epsilon''$$

where ϵ' is identical with the relative permittivity and ϵ'' is the dielectric loss.

As with mechanical moduli, this may be written in terms of the loss tangent tan δ, where

$$\tan \delta = \epsilon''/\epsilon'$$

This loss tangent corresponds to power dissipation within the dielectric [32,33].

4.5.1 *Definition of dielectric thermal analysis*

DETA is a technique in which the *dielectric constant* (permittivity, relative permittivity) and *dielectric loss factor* of the sample, in an oscillating electric field, are monitored against time or temperature while the temperature of the sample in a specified atmosphere is programmed. The instrument is a *dielectric thermal analyser*.

4.5.2 *Apparatus*

The furnace, programming and computer control used in DETA are similar to those in DMA and other thermal analysis techniques.

The sensors used in DETA may be parallel conducting plates between which the sample, in the form of a thin disc, is placed. A small force is exerted on the upper plate in order to improve contact with the sample. A temperature sensor may be included in one plate.

An alternative is to use a single surface with an array of electrodes. This may be placed onto the surface of the sample under slight pressure to obtain good contact. This single-surface sensor may be connected by a flexible lead for work remote from the main instrument [34].

It is also possible to adapt DSC or DTA systems to measure electrical resistance or permittivity [35,36].

4.5.3 *Frequency range*

The advantage of DETA is that a far wider range of frequencies can be used here than with mechanical perturbation in DMA, e.g. from about 1 Hz to 100 kHz, and several frequencies may be selected for measurement during an experiment by multiplexing.

This means that DETA is ideal for studying the kinetics of relaxation behaviour, as discussed above. High frequencies provide a fast data collection rate, most suitable for looking at fast changes, such as a rapidly curing polymer, where DMA would not be able to respond quickly enough.

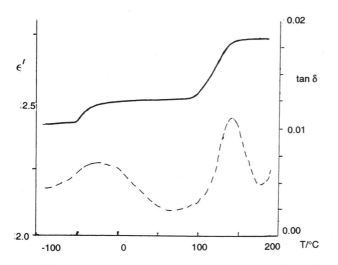

Figure 4.29   DETA curves for poly(ethylene terephalate). The dielectric constant increases with temperature and the tan δ curve shows two peaks, the strength of the process showing that the dipole of the ester group is active in this motion.

4.5.4   *Applications*

The applications of DETA are similar to those of DMA, with the advantages of an extended frequency range. It is important to recognise that for DETA to detect a response, the sample must contain, or have induced within it by the applied field, electrical dipoles. If these dipoles have a very small dipole moment, μ, then the effects observed will be small. The value of ϵ will vary during the relaxation process, from ϵ'_u at the beginning (unrelaxed) to ϵ'_R at the end (relaxed). The strength of the relaxation process $(\epsilon'_R - \epsilon'_u)$ may be correlated with the strength of the molecular dipole [33]:

$$\frac{(\epsilon'_R - \epsilon'_u)\,(2\epsilon'_R + \epsilon'_u)}{\epsilon'_R(\epsilon'_u + 2)} = \frac{4\pi g <\mu^2>}{3kT}$$

POLYMER TRANSITIONS

The DETA curve for PET over a wide temperature range is shown in Figure 4.29. The value of ϵ' increases with temperature, but the tan δ peaks show the alpha (glass transition) and beta peaks clearly. The strength of the beta process indicates that a large dipole, probably the ester grouping, is active in this motion.

The complementary nature of DMA and DETA is seen from Figure 4.30, which shows the same relaxation process for poly(ethylene terphthalate)

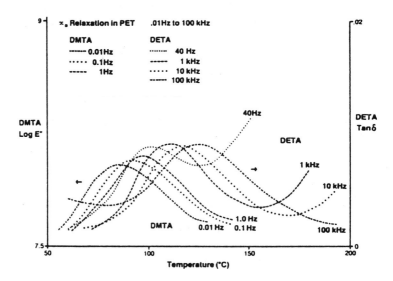

Figure 4.30 Comparison of the DMA and DETA results for PET. The low frequencies (0.01–1 Hz) measured mechanically transfer smoothly into the higher frequencies (40 Hz to 100 kHz) measured dielectrically (1 K/min) [32].

at frequencies from 0.01 to 1 Hz by DMA and from 40 Hz to 100 kHz by DETA, scanned in both cases at 1 K/min. The DETA curves show an upsweep at higher temperatures due to conductivity effects.

PHASE-SEPARATED MATERIALS [37]

The detection of the two phases in rubber toughened thermoplastic polymer can be seen by either technique as two peaks in the tan δ curve, one at low temperature (e.g. −70 °C) due to the dispersed rubber phase, the other at higher temperature (e.g. +100 °C) due to the glass transition of the thermoplastic.

Applications to coated materials, to natural fats and waxes and to polymer cure have also been reported.

It is also possible to make measurements simultaneously by using a remote DETA sensor embedded into the sample which is then clamped into the DMA system. This has the advantages of all simultaneous techniques, as outlined in Chapter 5, and allows a single experiment to determine both DETA and DMA parameters.

**4.6
Thermally stimulated
current analysis and
relaxation map analysis**

Thermally stimulated current analysis (TSC or TSCA) is a technique that involves heating the sample to a temperature above the transition of interest and then applying a static electric field E to induce polarisation P.

The sample is then cooled rapidly to a temperature considerably below the onset temperature of the transition so that the orientation caused by the polarisation is 'frozen' in place. The field is then cut off.

On reheating, the polarisation relaxes, generating a current which is measured by a sensitive electrometer and gives a 'dynamic conductivity' [38].

TSC and the related techniques of thermally stimulated luminescence and depolarisation (TSL and TSD) applied to organic polymers were reviewed by Fleming [39].

An interesting variant of this approach is called 'windowing'. Since transitions normally occur over several tens of degrees Celsius, the sample is heated to a temperature T_p within the transition range and the electric field is applied for a known time. The temperature is then reduced by 5–10 degrees to T_d, the field switched off and the sample held at that temperature for a short time. The sample is then quench cooled. In this way, only those motions corresponding to the narrow temperature interval (or 'window') traversed while the field was applied are frozen in an oriented manner. On reheating, the response measured is taken to arise *only* from that part of the broad spectrum of relaxations of which the transition is composed that were 'captured' by the windowing technique. By varying the values of the temperature 'window' the transition can be broken down into a large number of 'slices' which may be analysed separately and used to construct the *relaxation map analysis* (RMA) of the material [40]. In this way it is possible to gain a greater insight into the complexities of relaxation phenomena.

4.6.1 *Definition of thermally stimulated current analysis*

This is a technique that monitors, against time or temperature, the current that is generated when dipoles change their alignment in the sample, while the temperature of the sample in a specified atmosphere is programmed.

4.6.2. *Apparatus*

The sample is contained between two electrodes within a furnace, with cooling facility by nitrogen gas or liquid nitrogen, that can give cooling rates up to -60 K/min. The electric field is applied across the electrodes and subsequently the current of 10^{-17} to 10^{-8} A is measured.

4.6.3 *Applications*

Besides the complementary nature of this technique to the other thermo-analytical and thermoelectrical methods for the determination of glass

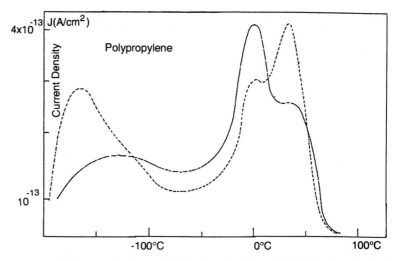

Figure 4.31   TSC curves for drawn (full line) and undrawn (dashed line) polypropylene, showing the polarisation recovery as the sample is heated [40].

transitions and beta and gamma transitions of polymers and composites and curing processes, TSC and RMA have been used to investigate the glassy state of amorphous polymers such as polystyrene, and the relaxation behaviour of different regions in semi-crystalline polymers. Figure 4.31 shows the TSC curves for drawn and undrawn polypropylene, and Figure 4.32 shows the RMA for the undrawn polypropylene. The relaxation component at lower temperatures is attributed to the inter-spherulitic regions, and that at higher temperatures to the chains whose ends are under constraint from the crystallites [40].

A study of PEEK [41] has shown the existence of two amorphous phases.

TSC has also been used for the investigation of bone [42]. Analysis of the fine structure of the complex TSC of bone and of products from the demineralisation of bone and comparison of the organic and mineral phases and the pure components have shown the importance of the interphase.

4.7
Problems
(Solutions on p. 276)

1. Suppose you wished to design a TMA system that used an optical sensor to detect the movement of the probe. What type of sensor might be suitable? Sketch the whole apparatus and mention any *disadvantages* of this type of system.

2. A metal rod 3.00 mm long expands by 7.5 μm when heated from 0 to 100 °C. Calculate the coefficient of linear expansion. If a block of the same metal measuring 3.0 × 5.0 × 10.0 cm was heated over the same

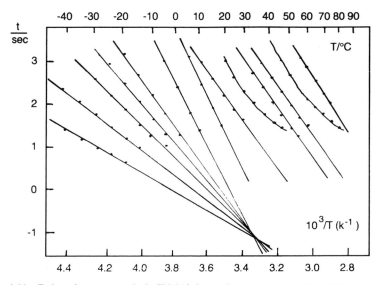

Figure 4.32 Relaxation map analysis (RMA) for undrawn polypropylene. The lines show three types of behaviour: straight Arrhenius, WLF curvature and lines obeying a compensation law. The lines converging in the low-temperature region may characterise a glassy phase under stress [40].

temperature range, how much would its volume increase and what is the coefficient of *volume* expansion?

3. A sample of polyamide film of thickness 1.00 mm at 30 °C expanded by 5 μm when heated to 80 °C, by a further 5 μm up to 120 °C and by 7.5 μm more up to 150 °C. Estimate the glass transition temperature of this polyamide.

4. A penetration probe with a hemispherical end of radius $R = 2$ mm rests on a rubber material held at 54 °C. With a load of 20 g, the penetration depth, d, was 0.113 mm. Given the approximate formula:

$$F = 1.778E \cdot R^{0.5} \cdot d^{1.5}$$

where F is the applied force, calculate the modulus E and compare it to the Young's modulus for this rubber of 2.3×10^6 Pa. What sort of TMA curve would this give?

5. Describe the difficulties associated with carrying out *constant strain* experiments on thin films and fibres. How are these difficulties overcome? Sketch the changes that would occur in the dynamic force as a fibre is heated through its glass transition. Illustrate how any other applied force would also change.

6. Two polymers are mixed together in solution and the solvent is then allowed to evaporate to form a solid. If this solid proves impossible to clamp in a DMA because it is too brittle, what alternative approach could you adopt? What would be the disadvantages of this approach?

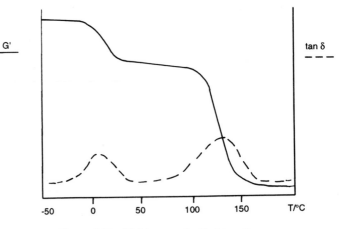

Figure 4.33   DMA curve for Problem 8.

How would you be able to tell if the polymers were miscible or immiscible?

7. Suggest ways of looking at the cross-linking behaviour of a polymer system that, before cross-linking occurs, is a low-viscosity liquid. What type of feature would you expect the DMA to show as the system cures? What important information about the properties of the fully cured system could you establish by DMA and how would you go about obtaining these data?

8. If you were given the DMA results shown in Figure 4.33 and are told that the sample consisted of only *one* polymer, how would you assign the peaks and the molecular motions from which they arise? If told that the sample consisted of two polymers, what *other* explanation might there be for the results?

9. For a particular sample of polycarbonate the storage modulus curves were determined at a series of temperatures and frequencies. In order to plot a 'master curve' referenced to 145 °C, the individual curves needed to be shifted along the log (frequency) axis by the amounts listed below.

Temperature (°C)	Shift factor ($\log(a)$)
135	3.36
140	1.57
145	0
150	−1.38
155	−2.60

(a) Show that these values may be fitted to a WLF equation with $C_1=22.9$ and $C_2=78.1$

(b) If the lowest measured frequency is 10^{-2} Hz, calculate the lowest frequency that could be found from the master curve.

10. A sample of polyisoprene which was originally a cube of side 1.65 mm was allowed to swell in toluene and the dimensional change measured by TMA. The final dimension was 2.83 mm. Given that the polymer–solvent interaction parameter $\chi_1 = 0.391$ for this combination and that the molar volume of toluene is 106.3 cm^3/mol, calculate the cross-link density using the Flory–Rehner equation.

References

1. J.O. Hill, *For Better Thermal Analysis and Calorimetry III*, ICTA, 1991.
2. F.W. Sears, M.W. Zemansky, H.D. Young, *College Physics* (5th edn), Addison-Wesley, Philippines, 1980.
3. F. Rodriguez, *Principles of Polymer Systems* (2nd edn), Ch. 8, McGraw-Hill, Singapore, 1983.
4. ASTM D4092–90, ASTM, Philadelphia.
5. B.K. Jones, *Electronics for Experimentation and Research*, Prentice-Hall, London, 1986.
6. TA Instruments Ltd, TMA 943 Brochure.
7. R.B. Prime, *Proc. 7th ICTA*, Wiley, Chichester, 1982, p. 984.
8. T. Ozawa, *J. Thermal Anal.*, 1993, **40**, 1379.
9. P.H. Willcocks, *J. Thermal Anal.*, 1993, **40**, 1451.
10. K.-H. Schuler. R. Sladek, P. Huse, *Thermochim. Acta*, 1988, **135**, 333.
11. ASTM D3386–84, ASTM, Philadelphia.
12. R. Riesen, W. Bartels, *Proc. 7th ICTA*, Wiley, Chichester, 1982, p. 1050.
13. W.P. Brennan, *Perkin-Elmer Application Study*, #26, 1978.
14. H.S. Yanai, W.J. Freund, O.L. Carter, *Thermochim. Acta*, 1972, **4**, 199.
15. R.B. Prime, in *Thermal Characterisation of Polymeric Materials* (ed. E.A. Turi), Academic Press, New York, 1981, p. 507.
16. W.P. Brennan, *Perkin–Elmer Application Study*, #23, 1977.
17. M.Y. Keating, *Dupont Application Brief*, TA-96, 1985.
18. Setaram, TMA 92, File 12: Applications Brochure.
19. M. El Sayed Ali, O.T. Sørensen, L. Hälldahl, *Proc. 7th ICTA*, Wiley, Chichester, 1982, p. 344.
20. E. Karmazin, M. Romand, M. Murat, *Proc. 2nd ESTA, Aberdeen*, Heyden, London, 1981, p. 562.
21. H.G. Wiedemann, M. Roessler, *Proc. 7th ICTA*, Wiley, Chichester, 1982, p. 1318.
22. R. Riesen, H. Sommerauer, *Amer. Lab.*, 1983, 30.
23. P.G. Fair, P.S. Gill, J.N. Leckenby, *Dupont Application Brief*, TA-90.
24. (a) R.E. Wetton, *Developments in Polymer Characterisation*, Applied Science, 1986, Ch. 5; (b) R.E. Wetton, J.S. Fisher, K.E. Pettitt, A. Evans, J.C. Duncan, *Amer. Lab.*, January, 1993, **25**, 16.
25. F. Biddlestone, A.A.S. Goodwin, J.N. Hay, G.A. Mouledons, *Polymer*, 1991, **32**, 3119.
26. M.L. Williams, R.F. Landel, J.D. Ferry, *J. Amer. Chem. Soc.*, 1955, **77**, 3701.
27. T.G. Fox, *Bull. Am. Phys. Soc.*, 1956, **1**(2), 123.
28. D.J. Lin, J.M. Ottino, E.L. Thomas, *Polym. Eng. Sci.*, 1985, **25**, 1155.
29. A.F. Lewis, J.K. Gilham, *J. Appl. Polym. Sci.*, 1962, **6**, 422.
30. M. Claybourn, M. Reading, *J. Appl. Polym. Sci.*, 1992, **44**, 565.
31. M. Reading, in *Thermal Analysis – Techniques and Applications* (ed. E.L. Charsley, S.B. Warrington), RSC, Cambridge, 1992.
32. R.E. Wetton, M.R. Morton, A.M. Rowe, *Amer. Lab.*, 1986, 70.
33. (a) H. Frohlich, *Theory of Dielectrics*, OUP, London, 1958; (b) V.V. Daniel, *Dielectric Relaxation*, Academic Press, New York, 1967.
34. S.T. Eckersley, A. Rudin, *J. Coat. Tech.*, 1990, **62**, 89.
35. A.K. Sircar, T.G. Lombard, J.L. Wells, *Thermochim. Acta*, 1980, **37**, 315.
36. K. Rajeshwar *et al.*, *Thermochim. Acta*, 1979, **33**, 157.
37. A.J. Mackinnon, D.J. Jenkins, P.T. McGrail, R.A. Pethrick, D. Hayward, C. Delides, A.S. Vatalis, *Macromolecules*, 1992, **25**, 3492.

38. A. Lamure, N. Hittini, C. Lacabanne, M.F. Herdmand, D. Herbage, *IEEE Trans. El.*, 1986, **21**, 443.
39. R.J. Fleming, *Thermochim. Acta*, 1988, **134**, 15.
40. M. Jarrigeon, B. Chabert, D. Chatain, C. Lacabanne, G. Nemoz, *J. Macromol. Sci., Phys.*, 1980, **b17**, 1.
41. M. Mourgues–Martin, A. Bernes, C. Lacabanne, *J. Thermal Anal.*, 1993, **40**, 697.
42. M. Mourgues, M.F. Harmand, A. Lamure, C. Lacabanne, *J. Thermal Anal.*, 1993, **40**, 863.

Bibliography   The general reference texts given in Chapter 1 should be consulted.

Simultaneous techniques and product analysis 5

P.J. Haines

**5.1
Introduction**

The previous chapters have concentrated on single techniques and their applications. However, we should remind ourselves that thermal methods often require complementary analysis by other techniques for a more complete understanding of the processes occurring. Even with the simplest decompositions – for example, when heating hydrated copper sulphate – we cannot be sure of the stages of reaction unless we confirm the products, both solid and gaseous, by other analyses. It would be useful to start by combining the thermal methods themselves, since they follow similar regimes of heating and atmosphere control, and should show the thermal events in a complementary manner.

The dynamic nature of thermal methods affects the results we obtain. Remembering SCRAM, we must consider the sample, the crucible or holder, the rate of heating, the surrounding atmosphere and the mass of sample. It would therefore be an advantage to use the same sample, the same heating rate and other conditions, but to sense the properties simultaneously. This saves *both* in time, since one run does the work of two or more, *and* on sample, if we have only a small amount to investigate. Most importantly, it gives results for two or more techniques under precisely the same experimental conditions.

It is only when two techniques are used on the same sample in the same run that the term 'simultaneous' should be used. When separate amounts of the same sample are run on different instruments, or even on different arrangements of the same instrument, then the term 'combined' or 'consecutive' should be used [1].

The products of thermal analysis may be analysed by any standard analytical method [2]. Gases can be separated by gas chromatographic columns and may be analysed by infrared or mass spectrometry, or by specific gas detectors, for instance. Solids can be examined by X-rays or by spectrometry, or taken through the regimes of separation and chemical analysis. It is also important to observe the nature of the change as it occurs, and running a sample on a heated stage under a microscope allows this to be done as well. As discussed above, it is a further advantage if these methods can be run simultaneously with the thermal analysis.

Ultimately, it might be possible to combine many analytical techniques, both conventional and thermal, together in a single, thermoanalytical system. One example of such a system was used in space exploration and produced data on the nature of the surface of Mars [3]. Uden and co-workers [4] have combined thermal analysers, pyrolysis furnaces, gas chromatography, vapour phase IR, mass spectrometry and elemental analysis into a single, complex system. They used the apparatus for the characterisation of oil shales, for polymers and organometallic compounds.

5.2
Simultaneous thermal analysis

Definition [1]   *Simultaneous techniques* refers to the application of two or more techniques to a sample at the same time. A dash is used to separate the abbreviations, e.g. simultaneous thermogravimetry and differential scanning calorimetry is TG–DSC.

5.2.1 *Simultaneous TG–DTA and TG–DSC*

As these are the most frequently used thermal methods, it is especially important to have a simultaneous technique incorporating both. The early work by the Pauliks and Erdey [5] used the derivatograph and allowed measurement of TG, DTG, DTA together. Many other designs have followed, and are often based on a thermobalance modified to weigh both sample and reference and to measure the temperature of each. A typical modern assembly for simultaneous TG–DSC (or 'STA') is shown in Figure 5.1.

The sample and reference materials are contained in crucibles, located on and supported by a heat-flux DSC plate. The leads to the temperature sensors below the crucibles go through the alumina support rod to the balance and detector systems. This system operates from −120 to 1500 °C, being cooled by a liquid nitrogen chamber for subambient operation. For temperatures from −160 to 2400 °C, a slightly different TG–DTA assembly may be used.

The SETARAM TG–DSC 111 uses symmetrical twin furnaces each with a Calvet microcalorimeter [7] while the TA Instruments SDT 2960 has a dual-beam horizontal design in which ceramic beams support the DTA sample and reference holders and dual meter movements are used to monitor the weight changes in sample and reference [8].

5.2.2 *Applications*

Many of the applications detailed in Chapters 2 and 3 may be carried out using the TG–DSC system. Some new examples will be used here.

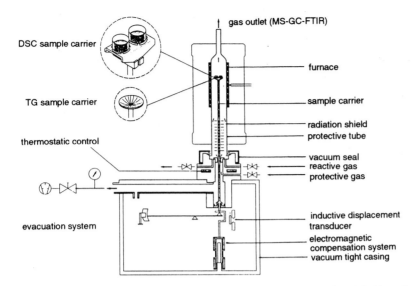

Figure 5.1 Netzsch STA 409 Simultaneous Thermal Analyser for TG–DSC. Note the facility for evolved gas analysis.

CALIBRATION MATERIALS [9]

Since it is quite difficult to calibrate the temperature with a conventional thermobalance system, and is much easier with DTA materials with well-characterised melting points, the use of simultaneous TG–DTA in a magnetic field allows the two calibrations to be performed in the same run. Figure 5.2 shows the melting endotherms of lead at 327 °C and of zinc at 419 °C which show only on the DTA, plus the Curie point transition of nickel which shows only on the TG. Thus, an accurate value for the temperature of this transition could be found as 358.1 ± 0.4 °C when a heating rate of 10 K/min was used.

SODIUM TUNGSTATE DIHYDRATE [8]

The decomposition of inorganic salt hydrates has been studied by single thermal analysis techniques, and the STA will give the TG and DTA information on the same sample at the same time. For example, the three-stage decomposition of calcium oxalate run on STA shows both the TG of Figure 2.15 and the DTA of Figure 3.35.

The dehydration of hydrated sodium tungstate $Na_2WO_4.2H_2O$ is shown in Figure 5.3. This material is used to standardise moisture analysis systems (see p. 170) and shows a single mass loss on the TG at around 100 °C of about 11% due to the loss of hydrate water, which should theoretically be

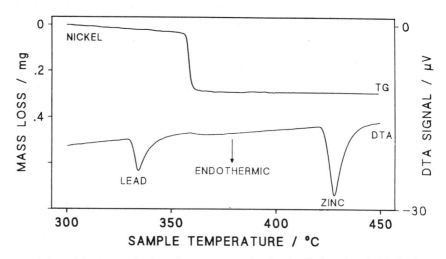

Figure 5.2   Determination of an accurate value for the Curie point of nickel using simultaneous TG–DTA in a magnetic field.

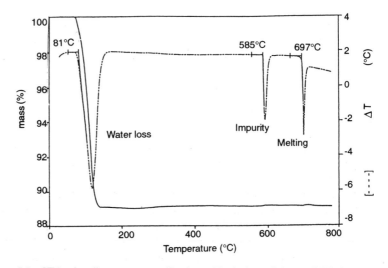

Figure 5.3   STA of sodium tungstate dihydrate. Typical conditions: 10 K/min, nitrogen at 100 cm³/min, 10 mg powdered sample.

10.9%. The DTA curve shows three endothermic events. Since the first coincides with the TG loss, it must represent the endothermic dehydration to anhydrous Na_2WO_4. The two sharp DTA endotherms with onsets at 585 and 697 °C must be phase transitions and are a crystal transition and the melting of the anhydrous sodium tungstate, respectively, in agreement with the literature.

More complex behaviour is shown by the decomposition of salts such as magnesium nitrate hexahydrate, $Mg(NO_3)_2.6H_2O$ [10]. This initially melts at about 90 °C and simultaneously starts to lose water. The DTA and TG curves are most uneven, since the sample loses more water with considerable bubbling. The final water is lost by 350 °C (42% loss) and the anhydrous salt melts at 390 °C before decomposing to the stable oxide (600 °C) leaving a residue of 15.7% of the original material. This complex behaviour needs additional techniques such as thermomicroscopy to sort out the stages.

FIRE-RETARDED WOOD [11]

The complex pattern of decomposition of wood is shown by the TG–DTA trace of Figure 5.4. The stages of combustion of cellulose have been identified [12,13] as dehydration and decomposition to laevoglucosan, followed and accompanied by production of flammable volatiles, tars, carbon and gases such as water vapour and oxides of carbon. Treatment of the materials with fire retardants such as borates, phosphates, or aluminium salts alters the temperatures and stages of decomposition. This may be observed very readily on TG–DTA. The small initial drying peak is little affected by the additive, but both the magnitudes and temperatures of the subsequent peaks are considerably changed. Tang and Neill [14] have compared the TG and DTA traces for a number of fire-retarded cellulose materials.

POLY(VINYL CHLORIDE) [15]

The thermal properties of polymers used as dielectric materials can be determined by STA. For example, filled PVC used as a cable sheathing gave the STA curves shown in Figure 5.5 in both nitrogen and air. The glass transition often shows on DTA at around 80 °C, and the dehydrochlorination reaction at around 270 °C is weakly endothermic, but shows a large weight change of over 50%. The subsequent loss of volatiles is very exothermic in air but less so in nitrogen. Small peaks are due to the additives present as fillers, fire retardants, heat and UV stabilisers, acid acceptors and char promoters [16]. The residue of around 10% in air comes from mineral additives and the larger residue in nitrogen is due to the uncombusted material and fillers such as carbon black.

PHARMACEUTICALS [17]

The nature of pharmaceutical materials, the water they contain and their thermal stability is often studied by TG–DSC. Figure 5.6 shows the loss of

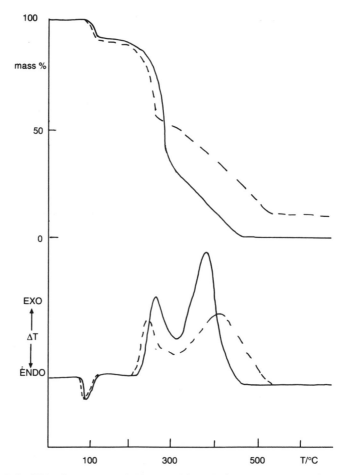

Figure 5.4   STA of poplar wood. Untreated (full line) and fire retarded (dashed line). Derivatograph trace after [11]; 10 K/min, air.

water from a hydrated pharmaceutical, followed by the melting of the anhydrous material at about 170 °C without decomposition, and lastly the start of an endothermic change above 250 °C.

REACTIVE ATMOSPHERE EFFECTS [7]

If the atmosphere above the sample is reactive, we may study the oxidation, reduction, catalytic and adsorptive reactions and corrosions. These all require an STA system that will not be adversely affected by any reactant or product materials. Figure 5.7 shows the reduction of a nickel oxide catalyst supported on silica by hydrogen. The water produced is detected by an attached mass spectrometer (see p. 171) and the three stages

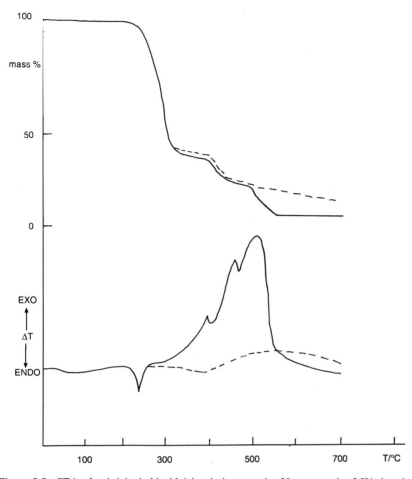

Figure 5.5 STA of poly(vinyl chloride) insulation sample; 20 mg sample, 5 K/min, air (full line) or nitrogen (dashed line) at 60 cm³/min (After [15]).

correspond to loss of adsorbed water around 100 °C and the reduction of the nickel oxide by the hydrogen around 500 °C. The DSC shows the endothermic nature of the desorption and the exothermic nature of the reduction.

5.3
Evolved gas analysis

In interpreting the mass losses in TG, or the peaks in DTA, we could not identify the products from that technique alone. Those thermal analysis techniques were not capable of differentiating between an endothermic loss due to solvent, or a decomposition reaction, or a physical process such as vaporisation. We very often need to identify the gases or vapours involved in the thermal process, but we must be most careful that they are *not* changed by further reaction or delayed by condensation or transfer lag.

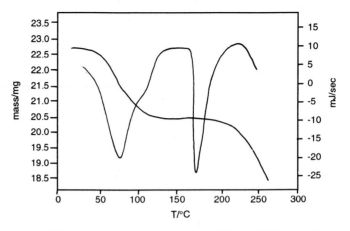

Figure 5.6 STA of pharmaceutical compound (22.6 mg, 10 K/min, nitrogen at 50 cm³/min).

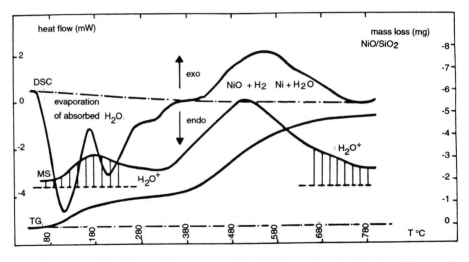

Figure 5.7 STA–EGA of the reduction of a NiO/SiO$_2$ catalyst in hydrogen; 30.1 mg, 10 K/min, 80% N$_2$, 20% H$_2$. *Note*: mass loss is here shown upwards. (Courtesy SETARAM).

Definition   Evolved gas analysis (EGA) is a technique in which the nature and/or the amount of gas or vapour evolved from the sample is monitored against time or temperature while the temperature of the sample in a specified atmosphere is programmed [18].

Where the gases are detected, but not analysed, the technique can be referred to as *evolved gas detection* (EGD).

5.3.1 *Instrumentation*

There are many ways in which the evolved gases may be detected and identified. For general thermal decomposition work, a large sample of several grams may be heated in a furnace in a flowing gas stream, using a controlled temperature programme. The products can be collected by passing the gas through an absorber, or through a cold trap, or through a series of specific gas detectors.

McNeill has developed the technique of *thermal volatilisation analysis* (TVA), where the sample is heated *in vacuo*, and the products are progressively condensed. The products are differentiated as the involatile residue, remaining in the oven, the high boilers, like waxes and less volatile oils, condensing on the walls, and the volatiles, passing on to the traps for condensation at lower temperatures [19,20]

For example, a copolymer of poly(vinyl bromide) (PVB) and poly-(methyl methacrylate) (PMMA), with 19% PVB, was shown to decompose in three stages by TG experiments. In a TVA experiment, the products from the first stage were volatile under vacuum at $-100\,°C$ and the product composition changed at $220\,°C$ and again at $325\,°C$. By separation, the products contained CO_2, HBr, CH_3Br, CH_3OH and MMA monomer. Infrared examination of the residue showed lactone rings.

5.3.2 *Apparatus*

It is an advantage to control the conditions by using a thermal analysis technique to provide the heating. Thermogravimetry has been widely used for this, and the gases evolved from the thermobalance are conducted away into other analytical devices. A typical arrangement is shown in Figure 5.8.

5.4 Detection and identification of evolved gases

We may classify the methods used for detection and/or identification under three headings:

1. Physical methods: conductivity, density.
2. Chemical methods: reaction, colour indication, electrochemical.
3. Spectroscopic methods: mass spectrometry, infrared.

We should add to this list the separation of mixtures of evolved materials by chromatographic techniques.

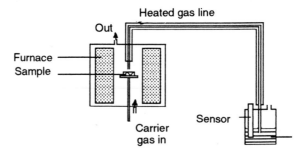

Figure 5.8 Schematic of TA–EGA system. The evolved gases are carried through the heated line by the carrier gas to the sensor (e.g. conductometric, pH, colorimetric) and/or the absorbent container.

5.4.1 *Physical methods*

The sensitive detectors of gases used for gas chromatography are often added to thermal analysis units to detect evolved gases. The DSC instrument could be fitted with a thermal conductivity detector to show changes in the gas streams. Flame-ionisation detectors (FID) have been used to detect gases evolved directly from heated plastic materials as well as gases separated by GC, but they will not, however, detect water vapour or carbon dioxide.

The gases evolved may be absorbed into a suitable solvent – for example, water vapour into dioxane or an organic vapour into an oil. The change can be followed by monitoring the electrical conductance or the capacitance of the solution or the solution may be analysed by chromatographic methods.

These methods often have the disadvantage that they are *non-specific*. Any gas which changes the physical properties will be detected.

In the *moisture evolution analyser* (MEA) [21], the moisture evolved from the heater is transferred by nitrogen carrier gas into the electrolytic cell detector where it is absorbed by phosphorus pentoxide coated onto platinum electrodes. The water is electrolysed to hydrogen and oxygen which are carried away by the gas stream. The electrolysis current is integrated and gives the amount of water directly. The system may be calibrated using water or a hydrate such as sodium tungstate dihydrate which has a theoretical water content of 10.92%. The technique forms the basis for the ASTM method D4019.

The capacitance moisture probe [22] has metal electrodes surrounding a hygroscopic dielectric layer which rapidly reaches equilibrium with the water vapour pressure, and changes the capacitance of the cell. A dew-point instrument has also been used to detect water [23]. There may be difficulties with these due to a slow, non-linear response to the moisture and lack of reversibility.

5.4.2 *Chemical methods*

The chemical reaction of gases with reagents or detectors specific to their chemical nature is a simple method of detecting single types of gases. The evolution of acidic gases, such as HCl from heated PVC, or of phthalic anhydride from polyester resins, can be detected and quantified by absorbing the evolved gas in a solvent such as aqueous alkali and measuring the change in pH, or the colour of an indicator or by eventual titration. For example, carbon dioxide evolved from heated concretes was dissolved and reacted with dilute, aqueous barium hydroxide and the reaction followed by conductance measurements [24].

Similar techniques can be used for the evolution of alkaline gases, such as ammonia, and McGhie *et al.* [25] report a simple pH method for measuring pH of a constantly renewed solution containing gases evolved from an ammonium/hydronium beta alumina:

$$(NH_4 \cdot H_3O)_{1.67} \, MgO_{0.67} \, Al_{10.33} \, O_{17}$$

The TG showed unresolved, multiple weight losses, but the pH trace clearly showed that most of the NH_3 is evolved below 400 °C, while the water evolved has little effect on the pH.

Many gases may be identified by the use of Draeger tubes. These contain reagents specific to particular gases, which change colour when they react. A semi-quantitative measurement of the gas concentration may be made.

Electrochemical detection using pH electrodes has been mentioned above, but electrodes sensitive to other species, such as metal ions, chloride, fluoride and cyanide ions, and gas-sensing membrane electrodes for such gases as CO_2, NO_2, H_2S, SO_2 and NH_3 are available [2]. The gas-sensing electrodes operate by establishing an equilibrium between the gas in the solution outside the electrode and the electrode membrane, which then equilibrates with the internal solution and internal electrode system.

A coulometric detector was used by Brinkworth *et al.* [26] to monitor sulphur dioxide evolution during coal combustion studies. The coal was burnt in a TG apparatus connected through an oxidation furnace to a coulometric cell. Any sulphur compounds evolved were oxidised to SO_2 which was then titrated [27] with iodine generated automatically. Figure 5.9 shows the weight loss and sulphur evolution profiles of three coals.

5.4.3 *Spectroscopic methods*

MASS SPECTROMETRY (MS) AND SIMULTANEOUS TG–MS

Mass spectrometry is a very good technique for identification of evolved gases and vapours. If a sample in the vapour state is introduced into the mass spectrometer under high vacuum, usually better than 10^{-5} torr, the

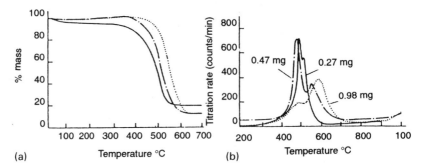

Figure 5.9   (a) Weight loss profiles and (b) sulphur evolution profiles for three coals heated in air: (i) high volatile C bituminous, 3.2% pyritic, 0.3% organic S (full line); (ii) semi-anthracitic, 0.1% pyritic, 0.6% organic S (dotted line); (iii) medium volatile bituminous, 1.0% pyritic, 0.7% organic S (dot-dash line) [26].

molecules M may then be ionised in various ways, for example by high energy electron impact using electrons accelerated by a potential difference of about 70 V:

$$M + e^* = M^{\cdot +} + 2e$$

The species $M^{\cdot +}$ is the molecular ion and generally a radical cation with an odd electron.

With such a high energy, there is a high probability that the molecular ion, $M^{\cdot +}$, may fragment into a lower mass fragment ion and a neutral fragment, for example, the molecular ion of propanone $[CH_3 \cdot CO \cdot CH_3]^{\cdot +}$ with a mass/charge ratio $m/z = 58$ readily fragments either to $CH_3 \cdot CO^+$ ($m/z = 43$) or to CH_3^+ ions ($m/z = 15$). The fragmentation pattern is characteristic of the molecule being studied.

Using other 'soft ionisation' techniques, such as chemical ionisation, fragmentation effects may be reduced.

The positive ions are accelerated through magnetic and/or electric fields which separate them according to their mass/charge ratios (m/z), and the ions are then detected by an electron multiplier [2].

The mass spectrometer operates in a high vacuum, which sets a problem of interfacing this spectrometer to any thermoanalytical system. The evolved gas is extracted from the thermal analysis instrument through a heated capillary leak at a suitable temperature. Methods such as jet or orifice separators have been used [28,29] to reduce the pressure before the mass spectrometer. The sampling capillary is set as close as possible to the sample and may be heated up to around 250 °C. A second stage reduction using a fine frit as a molecular leak may be added. This arrangement reduces the transfer time between the TG and the MS to less than 1 second [30,31,32]. A modern system is shown in Figure 5.10.

The software available with modern MS systems allows us to select

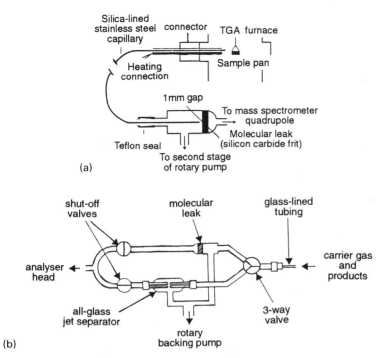

Figure 5.10   (a) TG–MS system; (b) dual inlet system for STA–MS.

several mass numbers and monitor them continously throughout the run, or to perform a rapid scan over a range of masses to obtain full mass spectra of the gases evolved. With proper materials and pumping there is little retention of the gases in the capillary or spectrometer, and so there is no 'memory' effect.

The 'total ion current' (TIC) shows the sum of all ionic species at a particular time, and may be compared to the thermal analysis curve.

CALCIUM OXALATE MONOHYDRATE (FIGURE 5.11)

This compound should be familiar by now! The TG trace shows three steps over the temperature range up to 100 °C, and by examining the evolution of the molecules of water ($m/z = 18$), or carbon monoxide ($m/z = 28$) and carbon dioxide ($m/z = 44$), we can say most certainly what the changes represent. The run was performed in argon ($m/z = 40$), which was subtracted from the spectra as background. The evolution of water corresponds exactly with the first DTG peak at 200 °C, the evolution of CO with the second at 480 °C, and of CO_2 with the third around 770 °C. Note that there is also some CO_2 at the second stage, possibly due to the disproportionation of CO [33], and that the $m/z = 28$ trace also peaks during the third stage, due to the fragmentation of the CO_2 ions.

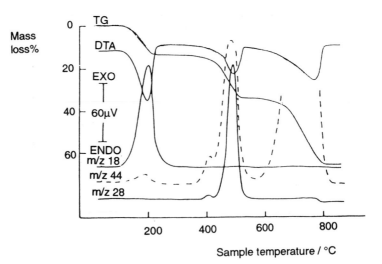

Figure 5.11   TG–DTA–MS curves for calcium oxalate monohydrate (15 mg, 15 K/min, argon).

POLY(ETHYLENE OXIDE) (FIGURE 5.12) [34]

Polymer decompositions may be extremely complex, and many products may be evolved simultaneously. The TG curves in air show a DTG peak with several overlapping parts. Monitoring of the prominent mass spectral peaks at $m/z = 73$ (probably a $C_3H_5O_2$ species) and $m/z = 88$ (p-dioxane, $C_4H_8O_2$) shows how they are evolved as a function of temperature. Different samples from different suppliers gave very different traces, and performing the experiment in nitrogen also altered the scans.

Studies on PVC [31,35] (see also p. 179) have shown that the stages of that decomposition may be monitored by TG–MS and that the first stage corresponds to evolution of HCl, low molecular weight hydrocarbons, especially benzene, and vinyl chloride monomer, while later stages produce higher molecular weight materials and finally a residue.

BRICK CLAYS [31]

A sample of an Oxford clay used in the manufacture of common bricks gives the TG curve shown in Figure 5.13 when heated at 20 K/min in air. The evolution of water ($m/z = 18$) occurs below 200 °C while at 300 °C combustion of the organic material produces several volatiles. Higher temperature processes involving the reactions of clays, carbonates and pyrites produce the complex patterns of evolution of CO_2 ($m/z = 44$) and SO_2 ($m/z = 64$).

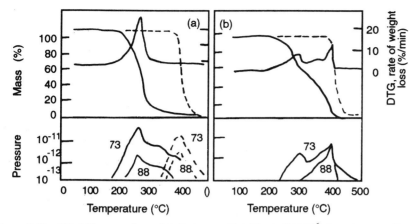

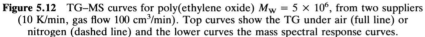

Figure 5.12   TG–MS curves for poly(ethylene oxide) $M_W = 5 \times 10^6$, from two suppliers (10 K/min, gas flow 100 cm³/min). Top curves show the TG under air (full line) or nitrogen (dashed line) and the lower curves the mass spectral response curves.

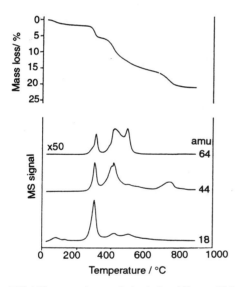

Figure 5.13   TG–MS curves for an Oxford clay (40 mg, 20 K/min, air) [31].

5.4.4   *Simultaneous thermal analysis – gas chromatography – mass spectrometry*

Although it is possible to obtain separate responses for several molecular species by thermal analysis – mass spectrometry, pyrolysis of samples such as polymers may give many products. It is an advantage if these can be

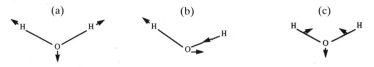

Figure 5.14   Vibrational motions of H_2O molecule. (a) Symmetric stretch 3652 cm^{-1}; (b) Antisymmetric stretch 3756 cm^{-1}; (c) Bending 1596 cm^{-1}.

separated between the thermal analysis unit and the mass spectrometer by gas chromatography. It is possible to do this by freezing the products in a cooled trap [36,37], but it is better to take the sample directly to avoid secondary reactions. Combined systems have been described [38,39] where this may be done, and important information has been obtained on the decomposition of polymer systems.

5.5
Infrared and
simultaneous
TA-infrared

Infrared spectrometry measures the absorption of light in the region of 2.5 to 50 μm (or 4000 to 200 cm^{-1}) due to its interaction with molecular vibrations. The number of fundamental frequencies of a molecule with n atoms is $(3n - 6)$, or, if linear, $(3n - 5)$. Not all of these may be active, since a condition for infrared absorption is that the molecular dipole must change during vibration.

Typical infrared bands with simple molecules involve the fundamental vibrations. For example, with the molecule H_2O, which is non-linear, the vibrations are shown in Figure 5.14.

In more complex molecules, the infrared spectrum is often interpreted in terms of the 'group frequencies', that is, vibrational frequencies typically observed for functional groups within the molecules. For example, in the molecule of propanone, $CH_3 \cdot CO \cdot CH_3$, we may assign the main infrared spectral bands as follows:

Group	Vibration	Wavenumber (cm^{-1})
C=O	stretching	1720
CH_3	stretchings	~2900
CH_3	bendings	~1400
C·CO·C	~bend	~1200

In gaseous samples, the molecular rotations also interact to give a complex 'ro-vibrational spectrum' with spectral bands. It should be noted that gaseous IR spectra often appear very different to the condensed phase spectra of the same chemical. For example, hydrogen-bonding effects are much less in the gas phase.

The spectra of many gaseous products may be measured by infrared spectrometry. The exceptions are homonuclear diatomic gases (e.g. N_2, O_2) whose molecules have no dipole. Although the spectra are often complex, there are advantages in using IR, particularly if a Fourier transform instrument is available. The infrared spectrum of a mixture of gases may be interpreted for each gas. The evolved gases may be carried from the furnace or thermal analysis system at atmospheric pressure in nitrogen or dry CO_2-free air or an inert gas. A long path-length through the gas is required, since the concentrations are low, and condensation of volatiles, especially water, should be avoided. Alternative methods condense the evolved material *either* in an argon matrix onto a cooled, reflective surface [40] which then moves into the path of the spectrometer *or* into a tube packed with a chromatographic absorbent such as Tenax for subsequent desorption, separation and detection.

5.5.1 *Apparatus*

While it is not intended to discuss infrared instrumentation fully here, it is essential to point out the main components and sampling methods used in infrared [2].

In a dispersive IR spectrometer, the infrared radiation from the IR source (e.g. a heated filament) is focused by mirror optics onto the sample. Since the container should absorb as little IR as possible, the sample is usually kept between windows of ionic solids, such as NaCl or KBr plates (or AgCl or CaF_2 plates if water may be present). The radiation is then dispersed by a grating or prism, and detected by thermocouple, pyroelectric or other detector.

Sample cells which may be heated (or cooled) are available and have been used to study the thermal behaviour and kinetics of reaction.

Fourier transform infrared (FTIR) spectrometers use similar sources, sampling devices and detectors, but work through the interference of two IR beams controlled by a mirror system. The interferogram is then converted by the Fourier transform software into a spectrum of transmittance against wavenumber. FTIR has several advantages over dispersive systems, particularly in time saving, in improved resolution and accuracy and in its ability to work over a greater range of infrared intensities.

Software allows the integration of the spectral intensity as a function of time to give the total evolution profile as shown by IR (Gram–Schmidt reconstruction).

Non-dispersive infrared analysers using an infrared source compare the IR intensity passing through a reference cell filled with a non-absorbing gas with that passing the sample cell through which the evolved gas flows. The detector cells are filled with the gas to be detected. The difference in absorption of IR relates to the concentration of IR-absorbing gas, e.g. CO_2. Their use for EGA has been discussed by Morgan [41] and also their application, with mass spectrometric and other EGA techniques to mineral sciences [42].

Reflection techniques, such as *attenuated total reflection* (ATR), where the sample is in contact with a prism of high refractive index, are useful and

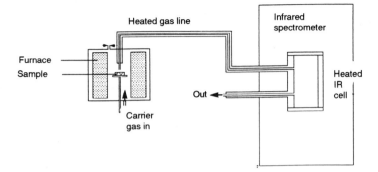

Figure 5.15 Schematic of a thermal analysis–infrared system.

minimise sample preparation. Thermal ATR units are now available so that the sample spectrum may be measured as a function of temperature. Diffuse and specular reflectance attachments have been adapted for heating experiments.

The general arrangement of a thermal analysis unit combined with infrared is shown in Figure 5.15.

5.5.2 *Applications*

AMMONIA EVOLUTION FROM MINERALS [43]

The evolution of gaseous ammonia from natural and synthetic ammonium-containing minerals has been carried out by non-dispersive IR. The minerals, in which the NH^+_4 ion has substituted for K^+, were heated in a furnace at 50 K/min and the gases carried over to the IR by a carrier gas stream flowing at 300 cm^3/min.

Figure 5.16 shows the EGA traces for ammonia and water evolved from NH_4-illite up to 1000 °C in both pure N_2 and mixed N_2/O_2 atmospheres.

The synthetic ammonium illite would have an ideal formula $NH_4Al_2(Si_3,Al)O_{10}(OH)_2$. The decomposition profile shows that there are two major losses, the first around 550 °C where *both* H_2O *and* NH_3 are evolved, and the second, around 750 °C, where the gas evolved is mostly H_2O.

This corresponds with the expected decompositions of the silicate-bound ammonium:

$$½[(NH_4)_2-O-] = NH_3 + 0.5H_2O$$

followed by the hydroxyl water release:

$$[Al_2(Si_3,Al)O_{10}(OH)_2] = [xAl_2O_3 + ySiO_2] + H_2O$$

A word of caution should be noted here. The amount of NH_3 detected

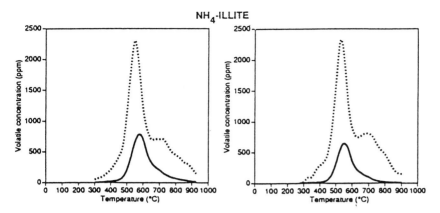

Figure 5.16 Evolved gas profiles for ammonium illite; 20 mg samples, in N_2 (left) and N_2/O_2 atmospheres. NH_3 (full line) and H_2O (dotted line) [43].

by integrating the EGA peaks is sometimes less than the theoretical, while the amount of H_2O is higher. It is likely that partial oxidation of the evolved NH_3 to nitrogen and water is responsible for this discrepancy.

POLY(VINYL CHLORIDE) (FIGURE 5.17 a–e)

The detection of HCl and other gases from heated PVC by TG–FTIR confirms the evolution of HCl and of benzene over the first peak of the DTG (Figure 5.17). The infrared spectrum in the region of the second peak suggests a complex mix of hydrocarbons and carbonyl compounds. The final weight loss around 800 °C gives a spectrum corresponding chiefly to carbon dioxide with some carbon monoxide. Since the run was performed in nitrogen, oxidation would be minimal, and the temperature, weight loss and residue suggest that there may be a considerable amount of inorganic filler, possibly calcium carbonate.

For a blend of copolymer of vinyl chloride and vinyl acetate, the TG–FTIR shows the evolution of HCl, of ethanoic acid (acetic acid) when heated in nitrogen and the production of CO_2 due to combustion in air at high temperatures (Figure 5.18).

POLYESTER RESINS

Polyester resins of the general formula:

$$-[-O-R_1-O-CO-R_2-CO-O-R_1-O-CO-R_3-CO-]_n-$$
$$\mid$$
$$[R_4]_m$$
$$\mid$$

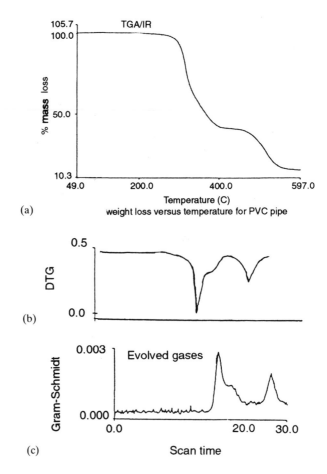

(a)   weight loss versus temperature for PVC pipe

(b)

(c)

Fig 17(a)–(c)

derived from a glycol HO–R$_1$–OH, an unsaturated diacid or R$_2$(COOH)$_2$, and a saturated or aromatic diacid R$_3$(COOH)$_2$ and cross-linked, generally with styrene (R$_4$), are widely used in construction, marine and engineering applications, especially when reinforced with glass fibre. Their thermal behaviour and fire-retardant properties are of great importance [45,46,47] and the gases evolved have been studied by trapping at low temperature, by mass spectrometry and pyrolysis.

In a study of a polyester resin prepared from a brominated glycol, maleic anhydride and phthalic anhydride and cross-linked with styrene, with 8% of a zinc stannate fire-retardant additive, the TG was run with analysis of the evolved gases by FTIR. The integrated (Gram–Schmidt) IR trace followed the TG and DTG curves closely, and the collected IR spectra showed the stages of decomposition. Figure 5.19 shows the TG and EGA (Gram–Schmidt) curves and Figure 5.20 shows part of the infrared spectral

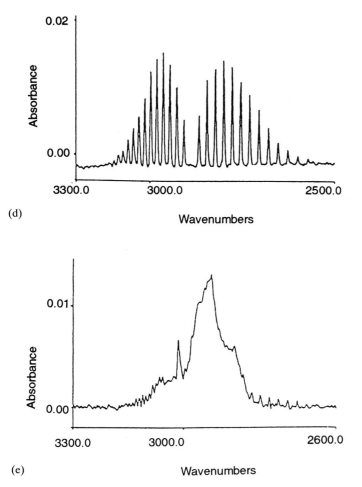

(d)

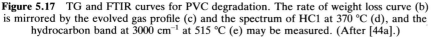

(e)

Figure 5.17 TG and FTIR curves for PVC degradation. The rate of weight loss curve (b) is mirrored by the evolved gas profile (c) and the spectrum of HCl at 370 °C (d), and the hydrocarbon band at 3000 cm^{-1} at 515 °C (e) may be measured. (After [44a].)

record. The peak at 2350 cm^{-1} is due to CO_2. At low temperatures (or times) an ester is evolved, probably methyl phthalate used in the catalyst. Between 200 and 320 °C the main product is phthalic anhydride, while above 320 °C styrene is evolved together with some HBr, CO, CO_2 and other products. It is also possible to select an IR 'window' and observe how the evolution of carbonyl compounds (peaks near 1700 cm^{-1}) changes during the analysis and to compare any IR spectrum with reference library spectra.

A study of the pyrolysis products from flame-retarded cotton and polyester–cotton fabrics used combined DTA–EGA infrared analysis to

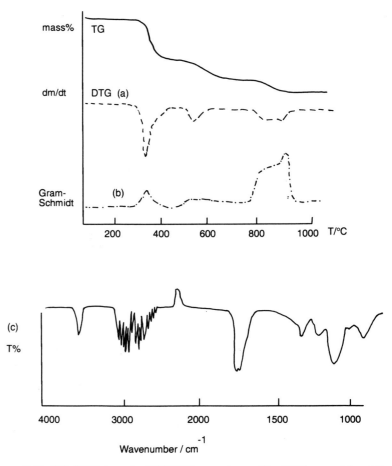

Figure 5.18 TG–FTIR trace for PVC/PVAc copolymer. The DTG curve (a) and the 'infrared thermogram' (b) show the events clearly. The infrared spectrum (c) of the gaseous products at the first TG loss (around 300 °C) shows the spectra of ethanoic acid (acetic acid) and the HCl band. (After [44b].)

study the profiles for evolution of CO, CO_2 and H_2O heated in nitrogen [48].

5.5.3 *Gas chromatography and pyrolysis GC–FTIR*

In many cases the gaseous products from heating real samples contain many chemical species: for example, on heating a polyester resin (Figure 5.21), up to 20 major peaks and 160 minor ones! It is sometimes possible to collect samples by trapping in a cold trap, or by adsorption on a suitable solid (e.g. 'Tenax') and to desorb onto a gas chromatographic column.

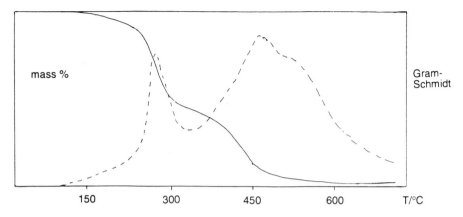

Figure 5.19   TG and Gram–Schmidt plot for degradation of a polyester resin; 20 mg, 25 K/min, nitrogen 100 cm³/min (TG, full line; Gram–Schmidt, dashed line).

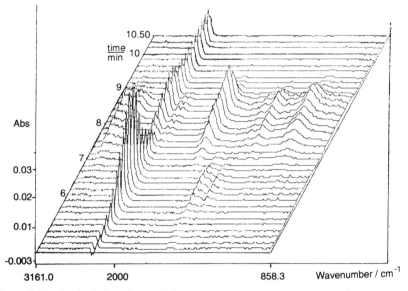

Figure 5.20   Stacked plot of part of the set of FTIR spectra recorded during the run of Figure 5.19.

Another alternative is to pyrolyse the sample for a few seconds at a selected temperature on a heated coil or ribbon and to sweep the products immediately onto a GC column using the carrier gas. By using a series of increasing temperatures the thermal degradation may be studied, although

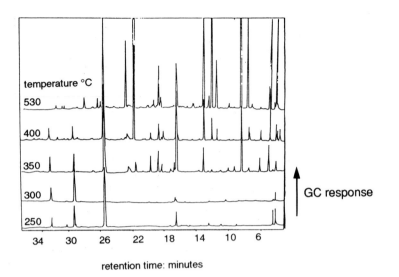

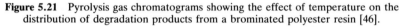

Figure 5.21 Pyrolysis gas chromatograms showing the effect of temperature on the distribution of degradation products from a brominated polyester resin [46].

it is necessary to point out that there is a very great difference between a retention time of a few seconds in a carrier gas stream and a time of several minutes under thermal analysis conditions. This method has been combined with FTIR and MS identification to give temperature-resolved separation of products or simulated thermal analysis. The books by Jones and Cramers [49], Irwin [50] and Vorhees [51] give many examples of the applications of pyrolysis techniques.

5.6
Infrared product analysis

The examples quoted above illustrate that IR may be of great use for evolved gas analysis. The products or intermediates of a reaction may be analysed by infrared spectrometry also. Modern FTIR instruments can work well in both transmission and reflection modes. Samples may be collected during a reaction and prepared for IR analysis in the usual ways in order to follow the course of the reaction. Alternatively, the IR spectra may be measured at various temperatures as the sample is heated in a thermally controlled IR cell using either transmission or reflection techniques.

5.6.1 *Application to polymer samples*

Mirabella [52,53] used an FTIR microscope to measure the transmission IR spectra of polymer samples simultaneously with their DSC curves. He

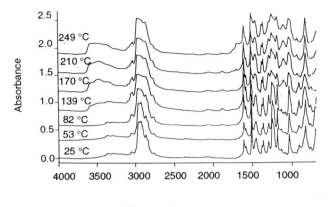

Figure 5.22   A series of transmission IR spectra at various temperatures during the heating of an epoxy sample containing 100 parts resin and 35 parts of activator [54].

showed that the IR intensities may change during heating through a physical change such as melting, and that degradation of a poly(vinyl alcohol) sample could be studied isothermally and showed a reduction in the hydroxyl band. Johnson *et al.* [54] measured the transmittance FTIR spectra at various temperatures for an epoxy resin system using a similar DSC–FTIR microscope system, and their results are shown in Figure 5.22.

Another technique is to reflect the IR from the surface of the sample in a metal DSC pan and record the reflectance spectrum, although under certain conditions this may give a reflectance IR spectrum that has a different appearance with distorted band shapes. However, this may be corrected using the Kramers–Kronig transformation [54].

The kinetics of polymerisation of epoxy resins have been studied by FTIR and the results compared with the DSC cure [55,56,57].

5.6.2 *Polymorphism*

Some endothermic changes detected by DSC could be due to a chemical or a phase change in the sample. In a comprehensive study of a pharmaceutical precursor, *p*-hexadecylaminobenzoic acid (HABA), Reffner and Ferrillo [58] used DSC, thermomicroscopy and FTIR microscopy. The first DSC trace showed endotherms at 90, 105 and 123 °C, but on second heating the first peak appeared at 71 °C. Thermomicroscopy and IR showed that there were three crystalline polymorphs, I, II and III. Form I changed to form III at 90 °C while form II changed at 71 °C, and form III melted to a liquid crystalline material at 104–105 °C. The liquid crystal became a true liquid at 123–124 °C. The IR spectra were measured at the temperatures at which each form was stable, and although the spectra were very similar, minor spectral differences could distinguish each phase.

5.6.3   *Metal complexes*

The changes that occur when metal complexes or metal organic compounds are heated are often followed by IR. The disappearance of characteristic IR bands, or the appearance of others, is a guide to the decomposition.

For example, in a study of the low-temperature synthesis of $BaBiO_3$ using citrate complexes of barium and bismuth nitrates [59], the IR spectra showed the ester formation around 185 °C, loss of organics and $BaCO_3$ formation by 380 °C and $BaBiO_3$ after prolonged heating at 850 °C. Similarly, the thermal decomposition of hydrated thallium(I) oxalates used TG, DTA, X-ray and infrared to clarify the stages in the decomposition [60].

5.7 Thermomicroscopy

It is often true that analysts may ignore the microscope as a tool, but it is also true that they may ignore heated stages and their methods [61]. It is frequently easy to identify the course of reactions and the nature of changes in physical properties by observation during heating.

Definition [18]   Thermoptometry is defined as 'a family of thermo-analytical techniques in which the optical property of the sample is monitored against time or temperature, while the temperature of the sample, in a specified atmosphere, is programmed'.

The observation of the sample under a microscope is *thermomicroscopy*, which will be our chief concern in this section. If the light intensity is measured, we can call it *thermophotometry* and if the method measures the light emitted from the sample, the technique is *thermoluminescence*.

It is very useful if the light intensity transmitted through the sample and system, or reflected from the sample, is monitored. Other refinements such as image analysis, video recording of the images, spectrometry and DSC have all been added to the basic hot-stage microscope, as shown in Figures 5.23 and 5.24.

The sample in the heated stage may be illuminated by plane polarised light using a polarising filter. If the sample is the same in all directions – for example, an isotropic liquid or a cubic crystal – it will not alter the polarisation.

Liquid crystals and solids of other crystalline systems will convert the light into elliptically polarised light due to differences in refractive index in the different directions and produce interference colours, characteristic of the sample. At a phase transition temperature, a change in the polarisation

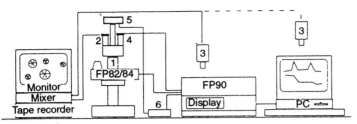

Figure 5.23 Mettler thermo-optical station including 1, microscope; 2 and 3, video cameras; 4, photomonitor; 5, camera; 6, handset [62].

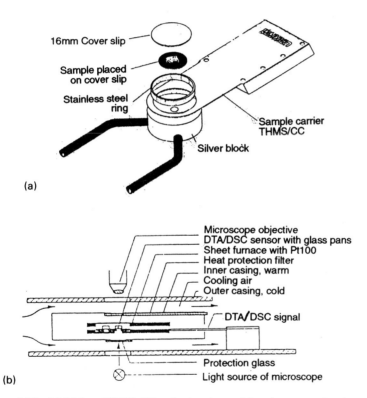

Figure 5.24 (a) Linkam THMS 600 series heating and freezing stage, showing sample preparation [63]; (b) Mettler FP 84 DSC hot stage [62].

will occur. The colours observed and the light intensity will also change at the transition. If the eyepiece has a polariser set at 90° to the polarising direction of the first filter ('crossed polars'), the changes may be seen most easily. Before melting, most crystals show as a regular, coloured pattern. After melting, the liquid appears black, since the light is completely cut off by the crossed polars.

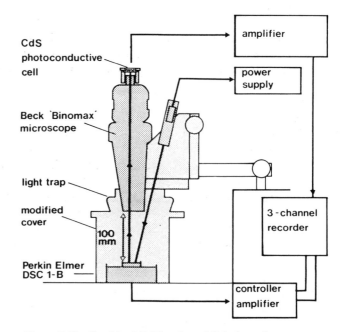

Figure 5.25 Combined DSC–reflected light intensity apparatus.

Thermomicroscopic equipment is available to work from $-180\,^{\circ}C$ to well above $2000\,^{\circ}C$. Polarised light hot-stage microscopy work is designated by several sets of initials: thermal depolarisation analysis (TDA) [64], de-polarised light intensity (DLI) [65], or thermo-optical analysis (TOA) [66].

Charsley *et al.* [67] have described a reflectance unit which has been used to study pyrotechnic reactions. Haines and Skinner [68] modified the Perkin–Elmer DSC 1-B to record the light reflected from the sample cell, as shown in Figure 5.25.

The use of thermomicroscopy simultaneously with DSC or DTA adds significantly to both techniques [69,70]. Rapid progress has been made in microspectrometry, and both Fourier transform IR (FTIR) and Raman microscopes incorporating heating or cooling stages are now widely used [71,72,73].

5.7.1 *Applications*

The applications of thermomicroscopy cross the entire range of materials from alloys, ceramics and minerals through foods, biomaterials and liquid crystals to pharmaceuticals, plastics and pyrotechnics. Examples will be chosen here to highlight the use of simultaneous thermomicroscopy–DSC/DTA.

5.7.2 *Phase equilibria*

The melting behaviour and purity of materials has been discussed in Chapter 3, but direct observation of the nature and extent of the melting may be made by thermomicroscopy. An estimate of the fraction melted may be made visually, or from measurements of the transmitted light trace, to complement the DSC measurements. Wiedemann and Bayer [74] describe an interesting application to azobenzene, which gradually transformed from the *cis*-isomer to the *trans*-isomer under illumination. Simultaneous thermomicroscopy–DSC has also been used to show that the double peaks on the DSC trace for tetratriacontane ($C_{34}H_{70}$) are due to a solid–solid transition at 69 °C followed by melting at 73 °C.

Barrall and Gallegos [75] used depolarised light intensity and DTA separately to record the phase behaviour of ammonium nitrate and polymers and compared the traces. Later workers have confirmed this behaviour using simultaneous measurements. The glass transition behaviour and stress relaxation of polymers and their blends has been studied by TOA and DSC [66] and considerable work has been done on the crystallisation rates of thermoplastic polymers [76,77]

Liquid crystals need to be studied by both DSC and thermomicroscopy, and the combined technique gives the greatest benefit [78]. A compound having a single nematic phase, such as *p*-azoxyanisole has two transitions. The crystal–nematic change gives a large DSC peak, but a small change in transmitted light (Figure 5.27). The nematic–isotropic change gives a small DSC peak, but a much larger optical change, as shown by the photograph in Figure 5.26.

Wiedemann [79] used this method to study the more complex phase behaviour of OOBPD (*N*,*N*'-bis(4-octyloxybenzylidene)-*p*-phenylenediamine) and Haines and Skinner [68] have shown that DSC with reflected light measurement can be used for phase studies.

Welch described the adaptation of the thermocouple as a sample holder, as well as a temperature sensor and heater [80], and Miller and Sommer [81] incorporated this principle into a DTA unit, using it to study the polymorphism of lithium and sodium sulphates.

High-temperature applications to inorganic compounds are described by Schultze [82] who recorded the DTA curve and simultaneously observed the congruent melting behaviour of $LiBi(PO_3)_4$ and the subsequent formation of a glass on cooling.

The application of thermomicroscopy to pharmaceuticals is discussed fully in an excellent, well-illustrated book by Kuhnert-Brandstatter [83]. Polymorphic changes within crystal systems mean that the crystal dimensions, types and structures change, and each different form interacts with polarised light in a characteristic way. A change from one crystal form to another is observed as a spread of altered shapes and colours of crystals across the field. Enantiotropic forms show distinct crystal habits. It is also

(a)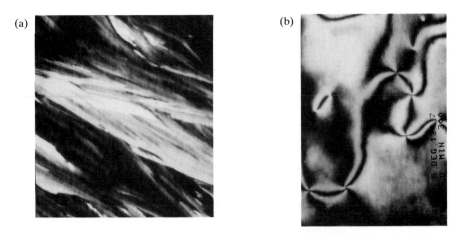 (b)

(c)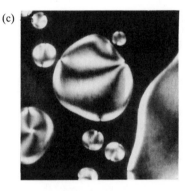

Figure 5.26 Polarised light microscope photographs of *p*-azoxyanisole heated to (a) 90 °C (crystal), (b) 130 °C (nematic) and (c) 135 °C (nematic going to isotropic liquid).

possible to investigate eutectic formation and refractive index by thermo-microscopic methods.

Wiedemann and Smykatz-Kloss [84] used the whole range of thermal analysis techniques (TG, DTA, TMA, TOA and MS) to study the phase transitions and losses of thenardite, Na_2SO_4, from various sources, and show the effects of sample composition, previous treatment and heating and the importance of thermomicroscopy in recognising the phase changes.

5.7.3 *Chemical reactions*

The light intensity transmitted or reflected by the sample may be recorded continuously during heating. This has the advantage, shared with TG, that

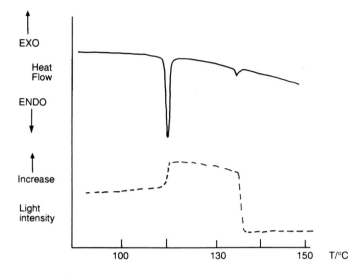

Figure 5.27 DSC (full line) and transmitted light intensity curves with crossed polars, for *p*-azoxyanisole (15 mg, 16 K/min, static air).

such changes are *additive*, and may indicate small, low-energy changes too small to register on DSC. In studies of reactions where the samples may be opaque, or may become so during reaction, there are advantages to using a reflected light system. Wendlandt describes a full system for dynamic reflectance spectroscopy [85] and surveys the applications, particularly to complex compounds. He points out that 'unlike TG or DTA, dynamic reflectance spectroscopy can be used to monitor only a single reaction at a time, thus eliminating the effect of reactions occurring simultaneously'.

5.7.4 *Hydrated salts*

Using simultaneous DSC and reflected light intensity (RLI), Haines and Skinner [68] showed how the decomposition behaviour of manganese formate dihydrate, $Mn(OOC.H)_2 \cdot 2H_2O$, may be followed. Figure 5.28 illustrates the changes recorded. Loss of hydrate water around 130 °C changed the crystals from translucent to opaque, and decomposition to manganese oxides, mostly brown MnO, occurs around 300 °C. The extra change shown by a decrease in RLI at 200 °C was shown by electron microscopy to be due to the cracking of the anhydrous formate.

Similar traces may be obtained for copper sulphate pentahydrate and for gypsum [86]. The dehydration of magnesium nitrate hexahydrate is more complex since initially the sample becomes liquid on heating to about 100 °C, and then the water bubbles off giving 'spiky' DTA and RLI traces,

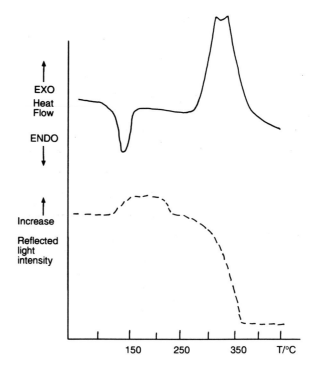

Figure 5.28   DSC (full line) and reflected light intensity curves for manganese formate dihydrate (5 mg sample, 16 K/min, static air).

until the anhydrous solid is formed around 250 °C. Fusion and decomposition to the oxide follow above 400 °C [87].

The ability to observe reactions and simultaneously to record the thermal traces has been important in the study of pyrotechnics [88], glasses and ceramics [89,90] and in elucidating the mechanism and kinetics of processes.

5.8
X-ray methods

The analysis of solid products may be carried out by chemical methods, but this does not reveal the structures – crystalline or otherwise – of the material. Polymorphic forms of the same material may give the same chemical analysis, but have very different structures and properties. X-ray analysis will reveal those structures and hence aid in the identification of the stages of reactions studied by thermal methods.

The diffraction of X-rays by a regular crystal lattice obeys the Bragg equation:

$$n\lambda = 2d \sin \theta$$

where n is an integer, λ is the wavelength of the X-rays, d is the spacing of the crystal planes and θ is the angle of diffraction.

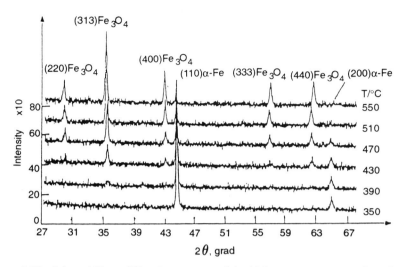

Figure 5.29 Selected X-ray diffraction patterns of the high-temperature corrosion of steel. Copper–Kα radiation.

The technique of X-ray powder diffraction [27] is particularly suited to the small samples used in thermal analysis techniques, and the powder diffraction patterns are indexed in a comprehensive set of tables [91]. Typically, the pattern is produced by the use of a collimated, monochromatic X-ray beam striking a powdered sample and producing a diffraction pattern, recorded photographically or electronically, which shows a series of sharp lines of characteristic intensity as a function of the angle θ, as shown in Figure 5.29.

The lines may be indexed according to the Miller indices of the crystal planes and/or the chemical species producing them, as shown in Figure 5.29. The X-ray pattern may be obtained using the reactant and products isolated after the reaction, or, as in Figure 5.29, the diffraction patterns can be obtained while the samples are heated and cooled in a programmed way. The X-ray diffractometer may be combined with a low- or high-temperature device to give a temperature resolved X-ray diffraction (TRXRD) unit [93].

Wide angle X-ray diffraction (WAXD) or wide angle X-ray scattering (WAXS) can be used in a similar way to powder diffraction to obtain a radially symmetrical, one-dimensional diffraction pattern from which the structure may be indexed according to Bragg's law.

X-rays are more strongly scattered by atoms of high atomic mass, and many applications involve inorganic reactions. Modern techniques also allow the study of organic and polymer systems. In small angle X-ray scattering (SAXS) the X-rays are scattered by regions with different electron densities. This provides a very good technique for studying the morphology of multiphase polymers, and it is often used in combination with DSC.

5.8.1   *Applications*

CORROSION OF STEEL [92]

The oxidation of steel in air is a complex solid/gas reaction, and layers of Fe_2O_3, Fe_3O_4 and FeO may grow on the Fe surface. The reactions should be studied as they occur and TRXRD gives the series of X-ray diffraction patterns shown in Figure 5.29. The patterns show that the main bulk of product at temperatures up to 550 °C is Fe_3O_4 and the formation of Fe_2O_3 and FeO is negligible. The kinetics of the reaction may be studied by TG and fit a parabolic oxidation law.

SYNTHESIS OF TOPAZ [94]

The reaction of Al_2O_3 and SiO_2 in the presence of ammonium fluoride will produce topaz $Al_2(SiO_4)(OH,F)_2$ under certain conditions. The thermal analysis curves for a mixture of corundum (natural Al_2O_3), quartz (SiO_2) and ammonium fluoride (NH_4F) in the ratio 1 : 1 : 6.4 are shown in Figure 5.30(a), and the X-ray diffraction patterns of the products at various temperatures are given in Figure 5.30(b). This complicated synthesis is shown to occur in at least four stages.

1. *About 150 °C:* Endothermic reaction of the corundum and quartz with the ammonium fluoride to produce complex fluorides:

$$Al_2O_3 + SiO_2 + 18NH_4F \rightarrow 2(NH_4)_3AlF_6 + (NH_4)_2SiF_6 + 10NH_3 + 5H_2O$$

2. *About 230 °C:* Endothermic decomposition of the complex aluminium fluoride:

$$(NH_4)_3AlF_6 \rightarrow NH_4AlF_4 + 2NH_3 + 2HF$$

3. *At 340 °C:* A sharp decomposition of the complex fluorides and volatilisation of products:

$$(NH_4)_2SiF_6 + NH_4AlF_4 \rightarrow SiF_4 + AlF_3 + 3NH_3 + 3HF$$

Figure 5.30   (a) Thermal analysis curves for the reaction of corundum–quartz–ammonium fluoride mixture of ratio 1:1:6.4. 500 mg, 10 K/min. (b)–(d) X-ray diffraction patterns of the products of the above reaction at (b) 155 °C, (c) 240 °C and (d) 800 °C. Cu–Kα radiation. A = NH_4AlF_4, B = $(NH_4)_2SiF_6$, C = $(NH_4)_3AlF_6$. T = topaz, Q = quartz, Cor = corundum.

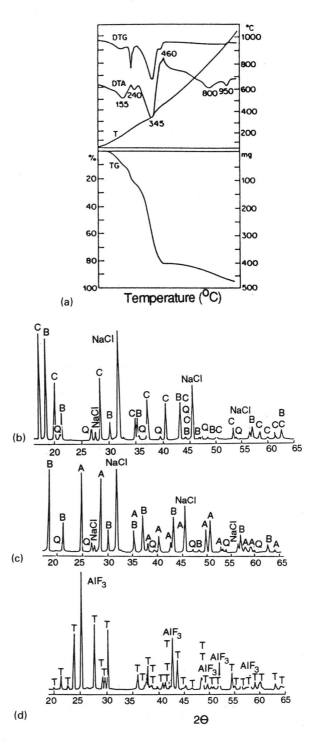

Fig 30(a)–(d)

(a)

Temperature (°C)

(b)

(c)

(d)

2θ

4. *At 800 °C:* A broad endotherm corresponding to the reaction of excess quartz to form topaz:

$$SiO_2 + 2AlF_3 + 4H_2O \rightarrow Al_2(SiO_4)(OH,F)_2 + 4HF + H_2$$

At higher temperatures, the topaz may dissociate to corundum and lose SiF_4.

FIRE-RETARDANT REACTIONS [95]

Fire retardants (FR) for polymers may act in several ways. The additive itself may be the chief fire retardant, for example by vaporising into the flame and quenching it or by undergoing endothermic decomposition to cool the reaction. Reactive monomers, such as halogen-containing diacids in polyester resins, may be both part of the polymer and also the chief producer of flame-quenching species. Other additives, particularly inorganic oxides such as Sb_2O_3 and MoO_3, may act by intermediate reactions with the polymer to produce less flammable products or flame-quenching species. Investigation of the behaviour of molybdenum trioxide in polyester resin systems showed that it acted as an excellent FR, especially in the presence of halogen-containing species. Thermal analysis showed that in air, the MoO_3 remained after degradation of the polymer, until it sublimed at around 750 °C. Evidence of reduction to MoO_2 was found in air above 300 °C, and in nitrogen the residual carbon in char reacted to give molybdenum carbide (see Figure 5.31).

POLYMER STRUCTURE

The powder diffraction pattern of polymers may be used in their identification; for example, the two crystalline forms of polyethylene can be distinguished by their patterns and polymer crystallinity can be estimated [96].

Conventional methods of obtaining small angle X-ray scattering (SAXS) and wide angle X-ray diffraction (WAXD) data may require long exposure times to obtain good results, and are limited to materials that are stable over this time scale. To get good X-ray patterns over a range of temperatures requires special apparatus capable of containing the sample and allowing passage of X-radiation. In order to combine X-ray methods with DSC a high flux of radiation and a fast detector must be used.

Koberstein and Russell [97,98] used synchrotron radiation as the X-ray source and a Mettler DSC modified to allow transmission of X-ray radiation plus a fast, position-sensitive electronic detector to obtain DSC/SAXS and DSC/WAXD results. They studied low-density polyethylene and thermoplastic copolyurethanes and found that there were semicrystalline segments and amorphous soft segments.

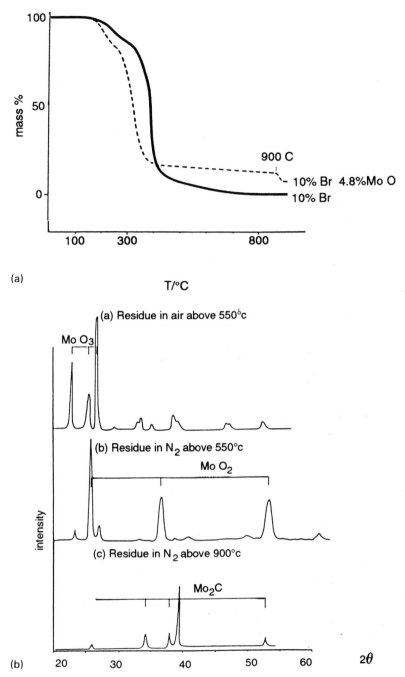

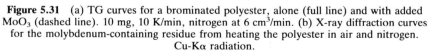

Figure 5.31 (a) TG curves for a brominated polyester, alone (full line) and with added MoO$_3$ (dashed line). 10 mg, 10 K/min, nitrogen at 6 cm^3/min. (b) X-ray diffraction curves for the molybdenum-containing residue from heating the polyester in air and nitrogen. Cu-Kα radiation.

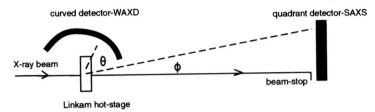

Figure 5.32   Schematic of the arrangement used for simultaneous SAXS and WAXD [99].

Ryan [99] has used synchrotron radiation monochromated to give a pinhole beamline of wavelength 0.152 nm and a quadrant detector for SAXS and a curved detector for WAXD covering 70° of arc, as shown in Figure 5.32. Recently, this has been combined with a DSC of single specimen design so that simultaneous SAXS–WAXD–DSC information may be obtained.

The measured scattered intensity $I(q)$ is a function of the scattering vector q, which has units of (length)$^{-1}$, and is defined by

$$q = (4\pi/\lambda)\cdot sin(\epsilon/2)$$

where ϵ is the scattering angle. The Lorentz correction may be applied to the intensity data because of the lamellar nature of a sample such as polyethylene to give the function Iq^2. The peak obtained on the curve of Iq^2 against q curve will give the domain spacing, d, from Bragg's law:

$$d = 2\pi/q_{peak}$$

The intensity may be converted to the invariant Q by the equation:

$$Q = \int I(q)\cdot q^2 \, dq$$

or to a 'relative invariant' from Q' from the area under the Iq^2 vs q curve between the first reliable data point ($q= 0.01$ Å^{-1}) to the region where Iq^2 becomes constant, that is at $q= 0.20$ Å^{-1}, as shown in Figure 5.33.

The WAXD patterns of high-density polyethylene (HDPE) are given in Figure 5.34 as a function of temperature. They show the sharp diffraction due to crystallites in the region 80–120 °C which broadens as the melting point is approached, until at 2 K above the melting point the pattern becomes that of an amorphous liquid.

The relative invariant Q' and its differential with respect to time (dQ'/dt) are shown in Figure 5.35 along with the DSC curve obtained simultaneously. The T_g is not observed in the SAXS patterns, but the temperature correlation of the crystallisation and melting events between the SAXS and DSC data is readily observed.

5.9
Electron microscopy and associated techniques

The very high magnifications possible with transmission electron microscopy (TEM) and the excellent depth of field that can be achieved with scanning electron microscopy (SEM) make them most useful tools for

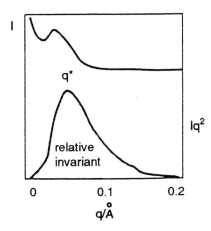

Figure 5.33 Plots of the intensity I and the Lorentz connected intensity Iq^2 for a semicrystalline sample of HDPE plotted versus the scattering vector q [99].

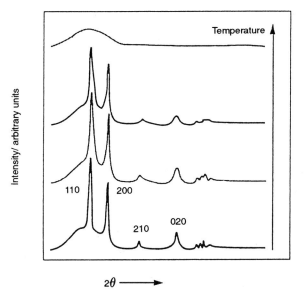

Figure 5.34 WAXD patterns of HDPE as a function of temperature. Numbers next to the lowest curve correspond to the Miller indices of the orthorhombic crystal [99].

examining the solid products from thermal analysis experiments. If the electron microscope is also equipped to analyse the X-ray spectrum produced from the sample by the electron impact – for example, by energy

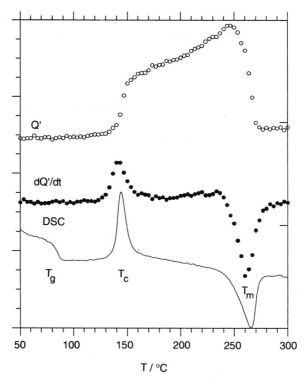

Figure 5.35 Combined SAXS patterns of the DSC for poly(ethylene terephthalate). The relative invariant Q' (open circles), its derivative with respect to time, dQ'/dt (filled circles) and the DSC plot are compared [100].

dispersive analysis of X-rays (EDAX) – then this will give an elemental analysis of the sample [2].

These techniques have been widely used as complementary methods to thermal analysis; for example, Galwey [101] has used SEM to characterise the kinetics and mechanism of solid-state reactions, and Reller *et al.* [102] have investigated the morphological changes in jadeite, $NaAl[Si_2O_6]$ and nephrite, $Ca_2(Mg,Fe)_5[(OH,F)Si_4O_{11}]_2$.

5.10
Conclusion   Thermal analysis techniques may be combined together with great advantages in time, sample and interpretation, but it is most necessary to use other analytical methods to confirm the identity of the intermediates and products that occur during the thermal processes.

5.11
Less common thermal   Several other techniques of thermal analysis are used to measure particular
analysis techniques   properties as a function of temperature or time. The definitions are given below and the references give a description or review of these methods and some of their applications, and additional information may be found in the

general reference texts listed at the end of Chapter 1. A special edition of *Thermochimica Acta* (1987, volume 110) was devoted to reviews of less common methods.

EMANATION THERMAL ANALYSIS (ETA)

A technique in which the release of trapped (usually radioactive) gas from the sample is monitored. [V. Balek, *Proc. 7th ICTA*, Wiley, Chichester, 1982, p. 371; *Thermochim. Acta*, 1987, **110**, 222.]

THERMOLUMINESCENCE (TL)

A technique in which the light (thermoluminescence) that is emitted from the sample is monitored. [T. Calderon *et al.*, *Thermochim. Acta*, 1988, **133**, 213.]

THERMOMAGNETOMETRY

A family of thermoanalytical techniqes in which a magnetic property of the sample is measured. [R. Moskalewicz, *Thermochim. Acta*, 1979, **28**, 229; S. St J. Warne, P.K. Gallagher, *Thermochim. Acta*, 1987, **110**, 269; P.K. Gallagher *et al.*, *J. Thermal Anal.*, 1993, **40**, 1423.]

THERMOSONIMETRY

A technique in which the sound emitted by the sample is monitored. [K. Lønvik, *Proc. 4th ICTA*, *Budapest*, 1974, Vol. 3, p. 1089; *Thermochim. Acta*, 1987, **110**, 253; S. Shimada, *J. Thermal Anal.*, 1993, **40**, 1063.]

5.12 Problems
(Solutions on p. 280)

1. You are presented with a VERY small sample of material and asked to characterise it. What methods would you choose? If thermal data were required, which *simultaneous* thermal methods might be most appropriate?
2. A sample of polymer was analysed by simultaneous TG–DTA. The thermal changes that might occur are: (a) glass transition; (b) plasticiser loss; (c) residual curing; (d) crystallisation; (e) melting; (f) oxidation; and (g) degradation. Which of these would be detected by DTA and which by TG?
3. A sample of an impure brick clay was heated in air to 900 °C and the evolved gases analysed by an infrared method. If the sample contained:
   (i) a clay $(Al_2O_3)_x \cdot (SiO_2)_y \cdot (H_2O)_z$

       (ii) gypsum, $CaSO_4 \cdot 2H_2O$

      (iii) calcite, $CaCO_3$

      (iv) organic matter and

      (v) pyrite, FeS_2

    account for the EGA peaks due to H_2O, CO_2, CO and SO_2.

4. A sample of calcium hydrazidocarbonate, $Ca(N_2H_3COO)_2 \cdot H_2O$, was heated in argon to 950 °C and the evolved gases studied by mass spectrometry. The losses corresponded as follows:

    (a) at about 100 °C to a peak in the $m/z = 18$ trace;

    (b) at about 350 °C to large peaks for $m/z = 17, 28$, and small peaks for $m/z = 44$ and 18;

    (c) at about 850 °C to peaks for $m/z = 28$ and 44.

    Suggest explanations for these observations.

5. During microscopic observation in a DSC run, a sample of a crystalline inorganic salt was found to undergo the following changes:

    (a) at 55 °C there was a colour change from blue to green and a sharp endotherm appeared on the DSC;

    (b) at 110 °C the crystals became opaque green and a broad DSC endotherm appeared peak;

    (c) at 180 °C the colour changed again to brown and a large DSC endotherm.

    Are these changes consistent with each other? Suggest an explanation for each observation. Would TG help?

6. Tristearin was studied by DSC in Chapter 3. A parallel thermomicroscopy study showed that using polarised light with crossed polars, the depolarised light intensity during heating

    (a) decreased nearly to zero around 56 °C then

    (b) increased to more than its initial value at 62 °C and

    (c) decreased again to zero at 75 °C.

    Explain these observations, using the comments in Chapter 3.

7. (a) The decomposition of zinc oxalate dihydrate, $ZnC_2O_4 \cdot 2H_2O$, was shown by TG and DTA to occur in two stages with a loss of mass of about 19% by 200 °C and a total loss of 57% by 400 °C. If we suggest the reactions as:

      (i) $ZnC_2O_4 \cdot 2H_2O \rightarrow ZnC_2O_4 + 2H_2O$ and

      (ii) $ZnC_2O_4 \rightarrow ZnO + CO + CO_2$

    how could we identify *both* gaseous *and* solid products?

    (b) An oxide gel formed by precipitation gave only a diffuse pattern on X-ray diffraction. On heating on a DTA it gave an exothermic peak at 640 °C and the product from heating gave a series of sharp lines on the XRD. Why?

References   1. G. Lombardi, *For Better Thermal Analysis*, ICTA, Rome, 1977.

2. F.W. Fifield, D. Kealey, *Principles and Practice of Analytical Chemistry* (3rd edn), Blackie, Glasgow, 1990.

3. M.I. Pope, M.D. Judd, *Differential Thermal Analysis*, Heyden, London, 1977.

4. P.C. Uden, D.E. Henderson, R.J. Lloyd, *Proc. 1st ESTA*, Heyden, London, 1976, p. 29.
5. F. Paulik, J. Paulik, L. Erdey, *Z. Anal. Chem.*, 1958, **160**, 241.
6. Netzsch STA 409 System Brochure.
7. P. Le Parlouer, *Thermochim. Acta*, 1987, **121**, 307.
8. C. Marcozzi, K. Reed, *Int. Lab.*, 1993, **23**(6), 10.
9. P.K. Gallagher *et al.*, *J. Thermal Anal.*, 1993, **40**, 1423.
10. Rheometrics PL-STA Brochure.
11. J. Simon, S. Gal, *Proc. 1st ESTA*, Heyden, London, 1976, p. 438.
12. F. Shafizadeh, *Adv. Carbohydr. Chem.*, 1968, **23**, 419.
13. D. Dollimore, J.M. Hoath, *Thermochim. Acta*, 1981, **45**, 87 and 103.
14. W.K. Tang, W.K. Neill, *J. Polym. Sci. (C)*, 1964, **6**, 65.
15. G. Stevens, *Fire & Materials*, 1983, **7**, 173.
16. J.J. Pitts, *Flame Retardancy of Polymeric Materials*, Vol. 1 (ed. W.C. Kuryla, A.J. Papa), Marcel Dekker, New York, 1973.
17. J. Hider, *Int. Labmate*, 1990, **XVII** (VI), 9.
18. J.O. Hill, *For Better Thermal Analysis and Calorimetry* (3rd edn), ICTAC, 1991.
19. I.C. McNeill, *J. Polym. Sci., Polym. Chem. Edn.*, 1977, **15**, 381.
20. I.C. McNeill, *European Polym. J.*, 1970, **6**, 373.
21. R.L. Hassel, *Amer. Lab.*, 1976, **8**, 33; and *Dupont Application Brief*, TA 46.
22. A.J. Diefenderfer, *Principles of Electronic Instrumentation*, Saunders, Philadelphia, 1979.
23. P.K. Gallagher, E.M. Gyorgy, W.R. Jones, *J. Thermal Anal.*, 1982, **23**, 185.
24. C.J. Keattch, *Analysis of Calcareous Materials*, SCI Monograph No. 18, London, 1964, p. 279.
25. A.R. McGhie, J.D. Denuzzio, G.C. Farrington, *Proc. 14th NATAS Conf.*, 1985.
26. S.J. Brinkworth, R.J. Howes, S.E. Mealor, *Proc. 2nd ESTA*, Heyden, London, 1981, p. 508.
27. D.A. Skoog, D.M. West, *Fundamentals of Analytical Chemistry*, Holt-Saunders, New York, 1982, p. 453.
28. E. Clarke, *Thermochim. Acta*, 1981, **51**, 7.
29. E. Kaisersberger, W.-D. Emmerich, *Proc 7th ICTA*, Wiley, Chichester, 1982, p. 279.
30. P.A. Barnes, *Proc. 1st ESTA*, Heyden, London, 1976, p. 31.
31. E.L. Charsley, C. Walker, S.B. Warrington, *J. Thermal Anal.*, 1993, **40**, 983.
32. E.L. Charsley, M.J. Manning, S.B. Warrington, *Stanton Redcroft Technical Information Paper*, TAMS/6.87/501.
33. D.H. Filsinger, D.B. Bourrie, *Proc. 7th ICTA*, Wiley, Chichester, 1982, p. 284.
34. E.L. Charsley, S.B. Warrington, G.K. Jones, A.R. McGhie, *Amer. Lab.*, 1990, **22**, 21.
35. C.F. Cullis, M.M. Hirschler, *The Combustion of Organic Polymers*, Clarendon Press, Oxford, 1981, p. 144.
36. J. Chiu, *Thermochim. Acta*, 1970, **1**, 231.
37. J.A.J. Jansen, W.E. Haas, H.G.M. Neutkens, A.J.H. Leenen, *Thermochim. Acta.*, 1988, **134**, 307.
38. L.F. Whiting, P.W. Langvardt, *Anal. Chem.*, 1984, **56**, 1755.
39. P.A. Barnes, G. Stevenson, S.B. Warrington, *Proc. 2nd ESTA*, Heyden, London, 1981, p. 47.
40. G.T. Ready, D.G. Ettinger, J.F. Schneider, S. Bourne, *Anal. Chem.*, 1985, **57**, 1602.
41. D.J. Morgan, *J. Thermal Anal.*, 1977, **12**, 245.
42. D.J. Morgan, S.B. Warrington, S. St J. Warne, *Thermochim. Acta*, 1988, **135**, 207.
43. S. Inglethorpe, D.J. Morgan, *J. Thermal Anal.*, 1993, **40**, 29.
44. (a). D.A. Compton, *Int. Labmate*, 1987, **XIV** (VI), 37; (b). B. Cassel, G. McClure, *Int. Lab.*, 1989 (April), 32.
45. P.J. Haines, T.J. Lever, G.A. Skinner, *Thermochim. Acta*, 1982, **59**, 331.
46. G.A. Skinner, P.J. Haines, T.J. Lever, *J. Appl. Polym. Sci.*, 1984, **29**, 763.
47. P.A. Moth, PhD Thesis, Kingston University, CNAA, 1992.
48. D. Price, A.R. Horrocks, M. Akalin, *Anal. Proc.*, 1990, **27**, 148.
49. C.E. Jones, C.A. Cramers (eds), *Analytical Pyrolysis*, Elsevier, Amsterdam, 1977.
50. W.J. Irwin, *Analytical Pyrolysis*, Marcel Dekker, New York, 1982.
51. K.J. Vorhees (ed.), *Analytical Pyrolysis*, Butterworth, London, 1984.
52. F.M. Mirabella, *Appl. Spectr.*, 1986, **40**, 417.

53. F.M. Mirabella, in *Infrared Microscopy* (R.G. Messerschmidt, M.A. Harthcock, eds), Dekker, New York, 1988, Ch. 6.
54. D.J. Johnson, D.A. Compton, P.L. Canale, *Thermochim. Acta,* 1992, **195**, 5.
55. M.K. Antoon, J.L. Koenig, *J. Polym. Sci., Polym. Chem. Edn*, 1981, **19**, 549.
56. R.J. Morgan, J.A. Happe, E.T. Mones, *28th National SAMPE Symposium*, 1983, p. 596.
57. J.M. Barton, *Adv. Polym. Sci.*, 1985, **72**, 111.
58. J.A. Reffner, R.G. Ferrillo, *J. Thermal Anal.*, 1988, **34**, 19.
59. G.K. Chuah, S. Jaenicke, K.S. Chan, S.T. Khor, J.O. Hill, *J. Thermal Anal.*, 1993, **40**, 1157.
60. S.R. Sagi, M.S. Prasada Rao, K.V. Ramana, *Proc. 7th ICTA*, Wiley, Chichester, 1982, p. 499.
61. W.C. McCrone, *Proc. 1st ESTA, Salford*, Heyden, London, 1976, p. 63.
62. H.G. Wiedemann, *J. Thermal Anal.*, 1993, **40**, 1031.
63. Linkam Scientific Instruments, THMS 600 Brochure.
64. G.W. Miller, *Analytical Calorimetry*, Vol I, Plenum, New York, 1970, p. 397.
65. E.M. Barrall, J.F. Johnson, *Appl. Polym. Symp.*, 1969, **8**, 191.
66. A.J. Kovacs, S.Y. Hobbs, *J. Appl. Polym. Sci.*, 1972, **16**, 301.
67. E.L. Charsley, A.C.F. Kamp, *Proc. 3rd ICTA, Davos*, Birkhauser-Verlag, Basel, 1972, **1**, 499.
68. P.J. Haines, G.A. Skinner, *Thermochim. Acta*, 1982, **59**, 343.
68. H.G. Wiedemann, G. Bayer, *Thermochim. Acta*, 1985, **85**, 271.
69. E.M. Barrall II, J.F. Johnson, *Thermochim. Acta*, 1972, **5**, 41.
70. G. Arneri, J.A. Sauer, *Thermochim. Acta*, 1976, **15**, 29.
71. D.A. Clark, G. Nichols, *Anal. Proc.*, 1990, **27**, 19.
72. M. Bowden, D.J. Gardiner, G. Rice, D.L. Gerrard, *J. Raman Spec.*, 1990, **21**, 37.
73. B. Cook, *Analytical Applications of Spectroscopy II* (A.M.C. Davies, C.S. Creaser, eds), RSC, Cambridge, 1991, p. 61.
74. H.G. Wiedemann, G. Bayer, *Thermochim. Acta*, 1985, **92**, 399.
75. E.M. Barrall, II, E.J. Gallegos, *J. Polym. Sci.*, A, 1967, **2**(5), 113.
76. F.L. Binsberger, B.G.M. de Lange, *Polymer*, 1970, **11**, 309.
77. C.F. Pratt, S.Y. Hobbs, *Polymer*, 1976, **17**, 12.
78. G.W. Gray, *Molecular Structure and Properties of Liquid Crystals*, Academic Press, London, 1962.
79. H.G. Wiedemann, Mettler Application No. 805.
80. J.H. Welch, *J. Sci. Inst.*, 1954, **31**, 458.
81. R.P. Miller, G. Sommer, *J. Sci. Inst.*, 1966, **43**, 293.
82. D. Schultze, *Thermochim. Acta*, 1979, **29**, 233.
83. M. Kuhnert-Brandstatter, *Thermomicroscopy in the Analysis of Pharmaceuticals*, Pergamon, Oxford, 1971.
84. H.G. Wiedemann, W. Smykatz-Kloss, *Thermochim. Acta*, 1981, **50**, 17.
85. W.W. Wendlandt (ed.), *Modern Aspects of Reflectance Spectroscopy*, Plenum, New York, 1968, p. 53.
86. K. Heide, *Thermochim. Acta*, 1972, **5**, 11.
87. E.L. Charsley, S.B. Warrington (eds), *Thermal Analysis–Techniques and Applications*, RSC, Cambridge, 1992.
88. (a) E.L. Charsley, D.E. Tolhurst, *Microscope*, 1975, **23**, 227; (b) G. Hussain, G.J. Rees, *Propellants, Explosives and Pyrotechnics*, 1991, **16**, 227.
89. G. Duma, *Sprechsaal*, 1986, **119**, 1035.
90. H.G. Ross, N. Malani, *J. Can. Ceram. Soc.*, 1979, **48**, 23.
91. *JCPDS Tables*, International Centre for Diffraction Studies, Pennsylvania, 1980.
92. W. Engel, N. Eisenreich, M. Alonso, V. Kolarik, *J. Thermal Anal.*, 1993, **40**, 1017.
93. A Deimling, W. Engel, N. Eisenreich, *J. Thermal Anal.*, 1992, **38**, 8430.
94. A.M.A. Rehim, *Proc. 7th ICTA*, Wiley, Chichester, 1982, p. 600.
95. P.J. Haines, T.J. Lever, G.A. Skinner, *Thermochim. Acta*, 1982, **59**, 331.
96. G.H. Edwards, *Brit. Polym. J.*, 1986, **18**, 88.
97. J.T. Koberstein, T.P. Russell, *J. Polym. Sci., Polym. Phys. Ed.*, 1985, **23**, 1109.
98. J.T. Koberstein, T.P. Russell, *Macromolecules*, 1992, **19**, 714.
99. A.J. Ryan, *J. Thermal Anal.*, 1993, **40**, 887.
100. A.J. Ryan, *ACS Symposium Series*, 1994 (in press).

101. A.K. Galwey, *Proc. 7th ICTA*, Wiley, Chichester, 1982, p. 38.
102. A. Reller, P.M. Wilde, H.G. Wiedemann, *J. Thermal Anal.*, 1993, **40**, 99.

The general reference texts given in Chapter 1 contain much material. One other text is very **Bibliography** relevant to this chapter:

W. Lodding (ed.), *Gas Effluent Analysis*, Arnold, London, 1967.

6 Problem solving and applications of thermal methods

Introduction   In industry and in academic research and teaching, thermal methods are used most frequently to solve problems, or to aid in their solution together with other analytical techniques. It is the aim of this chapter to illustrate typical problems and to suggest the thermal methods that may help in their solution or in the understanding of the phenomena that cause the problem. Many of the problems have arisen in the course of recent research in chemical and engineering laboratories. Some occurred in industry or consultancy where the problem had to be solved as quickly and as completely as possible.

Table 6.1   An overview of some of the papers presented at 10th ICTAC, Hatfield, England, August 1992

Environmental	
Agricultural residues	Ancient mortars
Carbonate minerals	Coal and other fuels
Conservation	Fly ash
Geological materials	Oil recovery
Zeolites	
Polymers	
Aromatic polymers	Bismaleimide resins
Copolymers	Degradation of polymers
Epoxy resins	Flame-retardant polymers
and many more!	
Pharmaceuticals and fine organics	
Dextran	Erythrocytes
Fatty acid esters	Liquid crystals
Organoplatinum antitumour agents	Proteins
New materials	
Carbide synthesis	Catalysts
Ferro-electric polymers	Lithium germanate glasses
Pyrochlores	Semi- and super-conductors
Thin films	

The range of application of thermal methods is now enormous! A survey of the papers presented at the 10th ICTAC Conference in 1992 shows how widely these techniques are now applied (Table 6.1).

In some cases, the thermal method has been well enough established that it is now used as a standard test method – for example, as one of the ASTM test methods. Several of these will be illustrated.

For comparison purposes, the problems have been grouped according to the materials being studied. The problems are presented in sequence and are followed by the suggested solutions. While these solutions are considered as the most probable answers in view of the evidence presently available, it is, of course, possible that later work will reveal that a different interpretation should be used. The original references should be consulted wherever possible.

The conditions used for the thermal analysis runs are stated in each case, and where a change in sample mass, atmosphere or rate of heating is known to make a difference to the interpretation, this is noted.

List of examples

Section 6.1   Inorganic materials
6.1.1   Tin(II) formate decomposition (TG, DTA, EGA, HSM)
6.1.2   Mixtures of carbonates (TG)
6.1.3   Strontium nitrate decomposition (TG–MS)
6.1.4   Calorimetry and phase transitions of potassium nitrate (DSC)
6.1.5   Decomposition of barium perchlorate (TG, DTA)
6.1.6   Solid-state reactions (DTA)

Section 6.2   Polymeric materials
6.2.1   Characterisation of a polymer (DSC, TG, DMA)
6.2.2   Polymer blend analysis (DSC)
6.2.3   Kinetic studies of polymer cure (ASTM 698)
6.2.4   Polymer decomposition studies (TG–MS)
6.2.5   Oxidative stability of polymers (DSC)
6.2.6   Studies of an epoxy–glass composite (DMA)
6.2.7   Characterisation of a thin adhesive film (DETA)

Section 6.3   Fine chemicals and pharmaceuticals
6.3.1   Purity determination (DSC)
6.3.2   Phase diagrams of organic chemicals (DSC)
6.3.3   Liquid crystal studies (DSC, TOA)
6.3.4   Stability and polymorphism of pharmaceuticals
6.3.5   Dynamic mechanical analysis of foods products

Problems

Caution In the solutions to all the following problems, and in the many other areas for which thermal methods have proved so useful, there are certain 'Cautionary Notes' that should be taken into consideration on every occasion.

1. The history of the samples and their *preparation* are very important. Ground samples, especially if they are mixtures, may behave very differently from 'chunky' samples.
2. The *run conditions* are most important for reproducibility and interpretation. SCRAM!
3. The *computing methods* used may be a great help, but we should also be aware of the pitfalls they can present:

   (a) A *noisy* trace may be a suggestion of instrumental problems which should be investigated. Excessive computer *smoothing* will not only remove the noise, but also any small signals!
   (b) Using the computer to increase the sensitivity may be overdone. The onset temperature for, say, an oxidation of an oil could be perceived as lower than before, simply because smaller changes are detected. A threshold value for the change might be a better approach.
   (c) Computers will do exactly what you tell them! If we propose that a certain change is an '*n*th order' reaction, the computer will sort out the data to give a best fit, and tell you that '$n = 4.321$' even though this may be a meaningless figure!

**6.1
Inorganic materials**

6.1.1 *Tin(II) formate decomposition*
(For solutions, see p. 245 and Reference 1)

Tin(II) formate (or tin methanoate) was prepared by refluxing tin(II) oxide with aqueous formic acid under a nitrogen atmosphere for several hours, and then crystallising the product. The salt is unstable and oxidises to give tin(IV) oxide in air and sunlight.

The salt was analysed by quantitative chemical analysis, infrared spectrometry and X-ray powder diffraction, and was shown to have the composition: $Sn(HCOO)_2$.

The decomposition was studied by TG and by DTA in the first instance and gave the traces shown in Figure 6.1.1.

TG and DTA

Sample: tin(II) formate, polycrystalline powder
Crucible: Pt dish

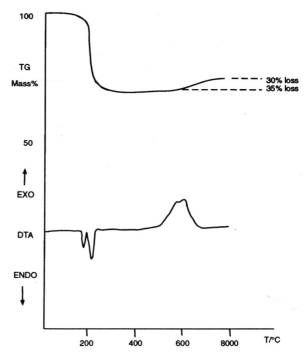

Figure 6.1.1   Thermal analysis curves for tin(II) formate.

Rate: 2–10 K/min
Atmosphere: static air
Mass: approx. 10 mg

Questions

(a) What are the stages of decomposition?
(b) Why does the DTA show *two* endotherms around 200 °C?
(c) How would you confirm the products of reaction?

6.1.2 *Mixtures of carbonates*
   (For solutions, see p. 246 and References 2–4)

Mixtures of metal carbonates can occur in nature, or as products of chemical reactions. Their decompositions are well characterised and distinct in temperature and mass loss. For mixtures of metal carbonates, is it possible accurately to measure the amount of each carbonate in a mixture?

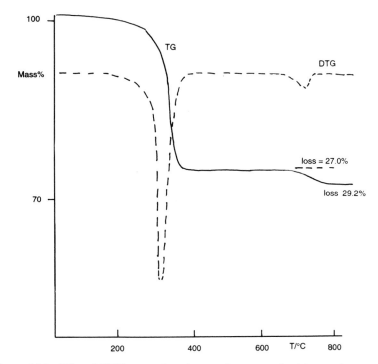

Figure 6.1.2   TG and DTG curves for mixture of copper and calcium carbonates.

An artificial mixture of calcium carbonate and 'copper carbonate' was made from laboratory reagent-grade chemicals by weighing and thorough mixing. The samples were analysed by TG only in flowing nitrogen (Figure 6.1.2).

TG

Samples: mixed calcium and copper carbonates, powder
Crucible: Pt dish
Rate: 10 K/min
Atmosphere: nitrogen, 30 cm^3/min
Mass: 10 mg

Questions

(a) The sample contained exactly 5.0% by weight of calcium carbonate. Is this confirmed by the TG trace?

(b) Several 'copper carbonates' are reported in the literature:

$$CuCO_3, \quad CuCO_3 \cdot Cu(OH)_2, \quad 2CuCO_3 \cdot Cu(OH)_2, \quad etc.$$

Which, if any, of these does the TG suggest is present?

(c) If *silver carbonate*, Ag_2CO_3, were added to the mixture, could all three components be detected accurately?

6.1.3 *Strontium nitrate decomposition*
(E.L. Charsley and S.B. Warrington, TACS, Leeds)
(For solutions, see p. 246 and References 5–8)

Strontium nitrate can be made by the usual methods, for example: strontium carbonate plus nitric acid, and also by double decomposition reaction of strontium chloride and sodium nitrate. It is used to produce the 'red fires' in pyrotechnics.

As part of a fundamental study of alkali metal and alkaline earth nitrates and nitrites and of pyrotechnic formulations containing them [5], a combined TG–MS study of the behaviour of strontium nitrate was performed.

The sample was shown to be $Sr(NO_3)_2$ by conventional analytical methods. Small sample masses and a helium atmosphere were used to remove the gaseous products efficiently from the reaction and prevent complications due to any reversible reactions.

TG–MS (Figure 6.1.3)

Sample: strontium nitrate, $Sr(NO_3)_2$, crystalline powder
Crucible: Pt boat

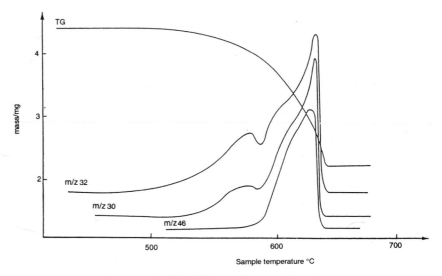

Figure 6.1.3 TG–MS curves for strontium nitrate.

Rate: 10 K/min
Atmosphere: He, 15 cm³/min
Mass: 4.4 mg
Interface: jet separator

Questions

(a) What do the TG and MS traces suggest as the most likely course of decomposition of the strontium nitrate?
(b) Is there any evidence for the formation of strontium *nitrite* $Sr(NO_2)_2$?
(c) What causes the dip in the mass spectrometric traces around 590 °C? How might you confirm your answer?

6.1.4   *Calorimetry and phase transitions of potassium nitrate*
(For solutions, see p. 247 and References 9–14)

In contrast to the previous examples, this is a problem in quantitative calorimetry. Potassium nitrate is easily obtained in high purity and is used as a standard ICTAC calibration material [9]. However, the phase behaviour is complex, and measurements of the thermodynamic parameters show a wide variation!

Potassium nitrate shows several phase changes [11], particularly a crystalline transition from Phase II (orthorhombic) to Phase I (rhombohedral) at 128 °C. On cooling, a third form (Phase III, also rhombohedral) appears.

A sample of ICTAC reference potassium nitrate, purified by recrystallisation and drying, was studied between the temperature limits 67 °C (340 K) and 197 °C (470 K) on a DSC. Calibration of the DSC was performed with a sapphire disc and with indium metal [10] and the temperatures were corrected. The traces are shown in Figure 6.1.4(a) for sapphire and for KNO_3.

DSC

Samples: (a) calibrant: sapphire disc, 24.08 mg
                    (b) KNO_3 powder, 33.11 mg
Crucibles: aluminium
Rate: 8 K/min
Atmosphere: static nitrogen

Questions

(a) Calculate the molar heat capacity of KNO_3 at at least two temperatures below the transition (Phase II) and at least two temperatures

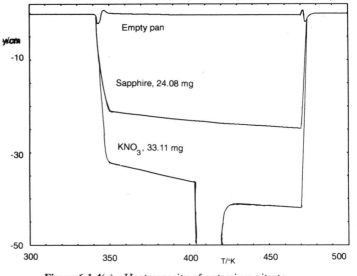

Figure 6.1.4(a)   Heat capacity of potassium nitrate.

above the transition (Phase I), given that the heat capacity of sapphire in this range may be calculated from the equation [14]

$$C_p \text{ (sapphire)/(J/(K g))} = 1.4571 - 3.355 \times 10^{-5} \times (T/K) - 200.17/(T/K)$$

(b) The value of the enthalpy change for the transition from crystalline form II to crystalline form I was found to be $\Delta H = 5.35$ kJ/mol and the corrected temperature of transition 401 K (128 °C). Calculate the *entropy change* ΔS for this transition.

Using the values of the molar heat capacity found in (a), correct the value of ΔH (transition) to the value at 298 K, and hence find the value for ΔH_f (I, 298 K) for the high-temperature form I, given that for the stable low-temperature form II,

$$\Delta H_f \text{ (II, 298 K)} = -497.1 \text{ kJ/mol}$$

What errors could affect this answer?

(c) Measurements of the enthalpy of transition of another crystalline form of KNO_3, form III to form I, have been made and it was found that ΔH (III–I) was 2.66 kJ/mol, about 400 K. The enthalpy of fusion of form I at 334 °C is about 11 kJ/mol. Estimate the enthalpy change for form III changing to form II at 400 K and the (hypothetical) enthalpy of fusion of form II.

Note   Values of the thermodynamic parameters show a considerable variation, e.g. values of ΔH (II–I) are quoted from 4.9 to 5.7 kJ/mol [10].

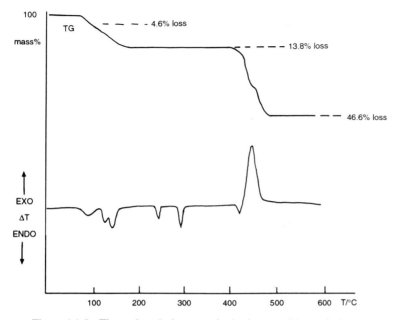

Figure 6.1.5 Thermal analysis curves for barium perchlorate hydrate.

6.1.5 *Decomposition of barium perchlorate*
(For solutions, see p. 249 and References 15–18)

The decomposition of barium perchlorate, $Ba(ClO_4)_2 \cdot 3H_2O$, was studied in order to see what effect semiconducting oxides (MnO_2 and Eu_2O_3) had on the kinetics and mechanism of the reactions [15].

The TG, DTG and DTA traces for the barium perchlorate trihydrate are shown in Figure 6.1.5.

Samples:	barium perchlorate powder
Crucible:	Pt dish or multiplate holder
Rate:	10 K/min
Atmosphere:	nitrogen, $10 cm^3/min$
Mass:	approx. 50 mg

Questions

(a) Explain, as far as possible, the features of the TG, DTG and DTA curves and suggest the nature of the products at both 300 °C and 600 °C.

(b) What other techniques could be used to confirm the identity of the products suggested above?

(c) When the pure hydrated barium perchlorate is mixed with MnO_2 in a molar ratio of 1:1, the thermoanalytical behaviour changes. Two molecules of water are lost first, then the final one. The temperature of the final decomposition is lowered from about 460 °C to about 310 °C and takes place in a single, smooth step. Suggest reasons for these effects.

6.1.6   Solid-state reactions
(For solutions, see p. 250 and References 19–23)

Mixtures of barium salts with potassium salts sometimes give very interesting thermal analysis curves. On cooling and re-running the products, the curve is sometimes quite different! The reaction between solid hydrated barium hydroxide and solid potassium nitrate [19] has been suggested as a means of solar energy storage.

$$Ba(OH)_2 \cdot 8H_2O + 2KNO_3 = Ba(NO_3)_2 + 2KOH + 8H_2O$$

The endothermic forward reaction around 80 °C ($\Delta H = 226$ J/g) is partially reversed on cooling below 60 °C but only gives out 138 J/g.

An investigation of the reaction of barium *chloride* and potassium nitrate gave the curves shown in Figure 6.1.6 [20,21].

Samples: solid $BaCl_2$ + solid KNO_3 in molar ratio 1 : 2
            large grains of mean radius = 0.5 mm
Crucible: platinum
Rate: 2 K/min

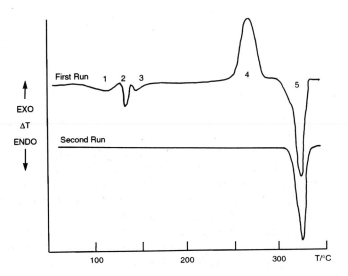

Figure 6.1.6   DTA curves for mixtures of 1 $BaCl_2$; 2 KNO_3.

Atmosphere: air, static
Mass: 0.5 g

Questions

(a) The TG shows small losses at $\approx 100\,°C$ on the first run, but not on re-running. Peaks 1–4 on the DTA also disappear when the product from heating to 350 °C is re-run, but Peak 5 still appears. Suggest an explanation. Would this reaction be suitable for solar energy storage?
(b) If the relatively large crystals used in this run are ground to a fine powder, Peaks 1–3 and 5 change very little, but Peak 4 is much sharper. Propose an explanation for this, and a method of testing.
(c) Peak 2 does not correspond to any change on the TG trace, nor does it appear in re-running the sample. Is this peak characteristic of either reactant?

**6.2
Polymeric materials**

6.2.1 *Characterisation of a polymer*
(For solutions, see p. 251 and References 24–30)

While spectroscopic methods of identification of polymers, especially infrared spectrometry, are extremely powerful and selective, with some samples it may be difficult or impossible to prepare a sample suitable for IR analysis, especially if that sample is highly filled!

The characteristic thermal behaviour of polymers does depend greatly on their nature, preparation and treatment. However, compilations of the typical properties of polymers are available [24] and we may measure many of these by thermal techniques.

The ASTM D3418 method [25] requires very specific treatment for the polymer in order to measure the glass transition temperature, melting temperature and crystallisation temperature. In summary, the method is as follows:

For a first-order transition

1. The sample is loaded into the pan, making sure there is good contact between the sample and the pan.
2. The sample is heated under nitrogen at 10 K/min to 30 °C above the melting point and held for 10 min to erase thermal history.
3. The sample is cooled to 50 °C below the peak crystallisation temperature at a rate of 10 K/min. The sample is next heated at 10 K/min and the heating (DSC or DTA) curve recorded.
4. After holding at the higher temperature for 10 min, the sample is cooled at 10 K/min and the freezing curve recorded. The temperatures T_f, T_m, T_c and T_e are recorded as shown in Figure 6.2.1(a) and a note is made if there are multiple meltings.

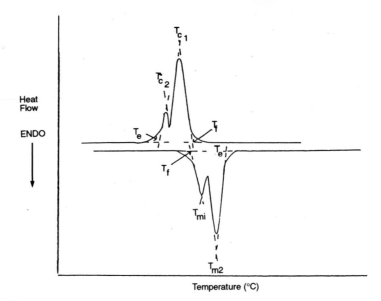

Figure 6.2.1(a)   ASTM definitions of melting and crystallisation temperatures.

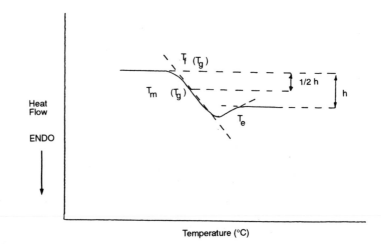

Figure 6.2.1(b)   ASTM definitions of glass transition temperatures.

For the glass transition

5. A fresh sample is taken and treated as in 1, 2 and 3 above, heating at 20 K/min. It is then quench cooled to 50 °C below the transition temperature and held for 10 min.

6. The sample is then heated at 20 K/min and the curve recorded. The temperatures shown in Figure 6.2.1(b) are reported.

The characterisation could be extended further by recording the TG curve of the material. This is the subject of ASTM D3850 [26], which relates to solid electrical insulating materials. Briefly, the standard test involves using a suitable calibrated thermobalance with a flow of dry air of 40–100 cm^3/min, heating at 5 K/min and recording the TG trace.

Characterisation using TMA and DMA is also very useful [27]. These will give additional information about the glass transition temperature and other low-temperature changes, plus the moduli of the polymer.

Sample

The sample provided was a white solid. This was prepared for analysis according to the directions given in the ASTM methods. A mass of about 10 mg was used for each technique and the samples were run in air at 10 cm^3/min (Figure 6.2.1(c)).

Questions

(a) Record the transition temperatures from the DTA trace as detailed above and identify the polymer present in the sample.
(b) Comment on the TG trace obtained.

6.2.2 Polymer blend analysis
(For solutions, see p. 251 and References 31–34)

A blend of polymers was submitted for analysis, and it was considered possible that the material contained a proportion of recycled polymer produced by 'Post Consumer Reclaim (PCR)'.

A very widely recycled polymer is poly(ethylene terephthalate), PET, used for the majority of soft drink plastic bottles. The major impurity in the recycled product would be the caps, made from polypropylene, PP. Many other bottles for household chemicals and shampoos are mainly high-density polyethylene, HDPE, and these too could contain PP as impurity.

The DSC traces for two samples of 'plastic' are given in Figure 6.2.2 together with typical thermal data for the constituents (Table 6.2.2). For *pure* PP, the melting peak at around 160 °C has an area of 100 J/g, under the same scanning conditions.

Samples:	plastic granules
Crucible:	aluminium
Rate:	10 K/min
Atmosphere:	nitrogen, 6 cm^3/min
Mass:	(a) 10.5 mg; (b) 17.1 mg

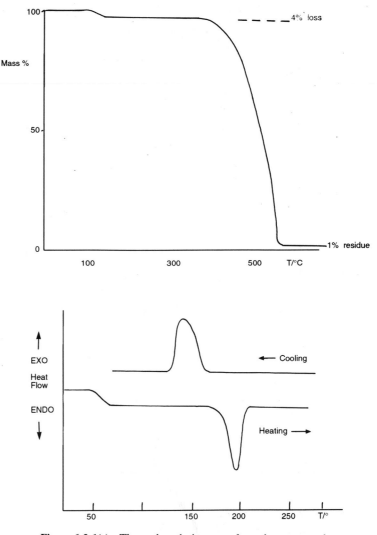

Figure 6.2.1(c)   Thermal analysis curves for unknown sample.

Questions

(a) Calculate the enthalpy of each peak in mJ/mg (or J/g) of mixture. Hence calculate the percentage of polypropylene in each sample and suggest what further thermal studies might be done to assist in quantifying the quality of the polymers for their end use.

(b) What further analysis would you suggest as confirmation of the composition of the blend?

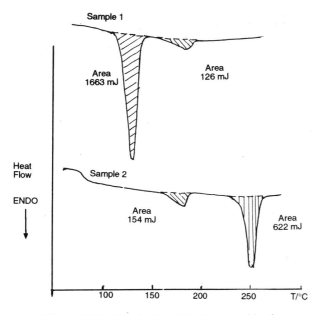

Figure 6.2.2   Samples 1 and 2 of polymer blends.

Table 6.2.2

Polymer	T_g(°C)	T_m(°C)	ΔH_m(J/g)
HDPE	−120	130	180
PP	−20	176	100
PET	80	250	40

6.2.3   Kinetic studies of polymer cure
(For solutions, see p. 252 and References 35–38)

Epoxy resins are used for many purposes, for example as matrix resins for advanced aerospace composites. It is important that the curing process and the relationship of resin structure to properties is well understood.

The sample studied here is a trifunctional epoxy compound:

$$[CH_2\!-\!CH\!-\!CH_2\!-\!O\!-\!C_6H_4]_3\cdot CH \qquad (XD)$$
$$\diagdown\!\!\diagup$$
$$O$$

together with a hardener, 4,4′-diaminodiphenylsulphone (DDS).

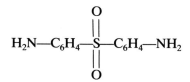

Calorimetric curing experiments were paralleled by both IR and rheological measurements. The fraction reacted was calculated from the infrared peaks of the epoxy and amine groups.

To investigate the most appropriate kinetic equation and $f(\alpha)$, isothermal runs were conducted at a series of temperatures between 125 and 225 °C. The product after a long cure was re-run to obtain a baseline correction. A typical *corrected* isothermal run is shown in Figure 6.2.3(a) and the data for one run are given in Table 6.2.3(a).

The data may then be used to interpret scanning DSC runs, as shown in Figure 6.2.3(b). Data from runs at several different scan rates are given in Table 6.2.3(b).

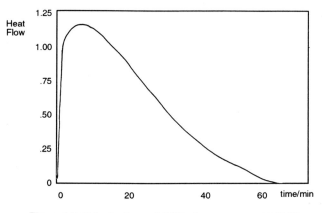

Figure 6.2.3(a)   Isothermal DSC of epoxy cure at 150 °C.

Table 6.2.3(a)

Time(s)	$10^4 \times$ rate	α
90	4.87	0.05
180	5.35	0.10
300	5.51	0.15
390	5.57	0.20
480	5.57	0.25
582	5.46	0.30
660	5.20	0.35
750	4.92	0.40
960	4.20	0.50

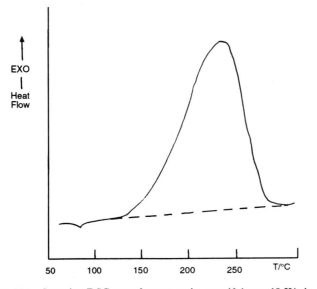

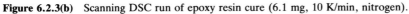

Figure 6.2.3(b)   Scanning DSC run of epoxy resin cure (6.1 mg, 10 K/min, nitrogen).

Table 6.2.3(b)

Rate (K/min)	Peak temperature (K)
1	431.3
5	467.9
10	485.7
20	504.9

Sample: XD–DDS
Crucible: aluminium
Rate   (a)0 (isothermal at 150 °C); (b) 10 K/min
Atmosphere: nitrogen, 30 cm^3/min
Mass: about 6 mg

Questions

(a) An equation used by several workers to describe cure kinetics [35,36] is:

$$\text{Rate} = (K_1 + K_2 \cdot \alpha^m) \cdot (1 - \alpha)^n$$

By plotting $[\text{Rate}/(1 - \alpha)^n]$ against α^m, we can obtain K_1 and K_2. Try this with $m = 1$ and $n = 2$ and $m = 1$, $n = 1$ (autocatalytic reaction).

(b) Ozawa [38] has shown that for runs at different scan rates β, for the same fraction α reacted:

$$\log_{10} \beta = \text{constant} - 0.4567E/RT$$

Plotting $\log_{10} \beta$ against $1/T_\alpha$ gives an approximate value of E. Corrections can be applied to refine the value of E, as shown in the ASTM 698. Calculate E.

6.2.4 *Polymer decomposition studies*
(E.L. Charsley and S.B. Warrington, TACS, Leeds)
(For solutions, see p. 253 and References 39–43)

The curing and subsequent degradation of phenol–formaldehyde novolaks cross-linked with hexamethylene tetramine (hexamine, HMT) is important and affects the properties of the finished products.

1. Novolaks are commonly prepared [39] by the reaction under acidic conditions of a molar excess of phenol with formaldehyde (methanal), usually about 1.25 : 1. The initial intermediate methylol phenols react rapidly to form linked, substituted phenols with the elimination of water. A general structure for this product is shown in **I**.

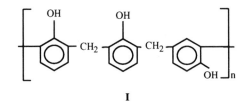

I

2. Novolaks may be converted into cross-linked networks by the further reaction with an agent, most often hexamethylene tetramine, HMT (see **II**).

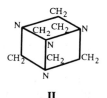

II

3. The reaction of HMT with phenols alone has been studied [40] and two crystalline products characterised by X-ray analysis. The curing and

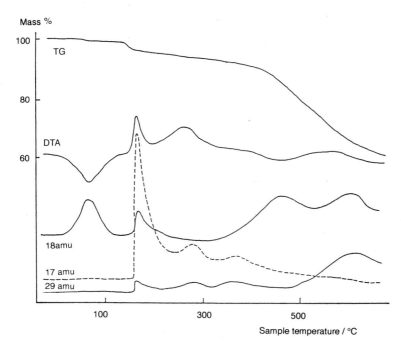

Figure 6.2.4   STA–EGA curves for phenol–formaldehyde–HMT.

degradation of novolak–HMT systems was investigated by TG, DTA and mass-spectrometric EGA [40,41] and the curves are shown in Figure 6.2.4.

Sample: phenol–formaldehyde novolak–HMT resin
Crucible: platinum
Rate: 10 K/min
Atmosphere: argon
Mass: 15 mg
EGA: mass spectrometric system

Questions

(a) Explain as far as possible the mechanisms of curing and degradation of the resin as shown by these curves.
(b) How might the cure kinetics be followed?
(c) If a fully cured and moulded sample of a filled novolak–HMT resin was run, what type of TG, DTA and EGA traces would you expect to obtain?

6.2.5 *Oxidative stability of polymers*
(For solutions, see p. 254 and References 44–47)

The widespread use of polyolefins for sheeting, for bottle caps, for electrical insulation and for plastic water piping requires a method of testing the thermal and oxidative stability of the polymer material. These tests are specified in the ASTM methods D3350–84 for polyethylene plastic pipes and fittings, and in D3895–80 for copper-induced oxidation of polyolefins [44].

The problem here concerns the degradation of water piping under natural conditions. The surface of a pipe with an outside diameter of 180 mm, coloured blue for coding, appeared to have deteriorated after use under normal conditions for 9 months [46].

There are two methods of testing of the oxidative stability of polyolefins by DSC: either scanning in air, or isothermal in oxygen. Both were used for these samples [45].

(a) Oxidation induction temperature

Samples of about 5 mg were cut, without using heat, from the inner, outer, surfaces of the pipe. These were placed in unlidded aluminium pans in the DSC. As a check a small sample of pure indium was used as calibration.

Sample: polyethylene blue water pipe
                (i) outer; (ii) inner.
Crucible: aluminium
Rate: 10 K/min
Atmosphere: oxygen, 50 cm^3/min
Mass: ≈5 mg

(b) Oxidation induction time (OIT)

Larger samples weighing about 15 mg obtained as above were placed in the DSC and heated in nitrogen to 200 °C. After equilibrating for 5 min, the gas was switched to oxygen and the timing started.

Sample: polyethylene blue water pipe
                (iii) outer; (iv) inner
Crucible: aluminium
Rate: isothermal at 200 °C
Atmosphere: nitrogen, 50 cm^3/min then oxygen, 50 cm^3/min
Mass: ≈ 15 mg

Questions

(a) Compare the stability of the samples as measured by each test.

(b) Suggest what other tests or techniques might help to investigate this problem.

6.2.6   *Studies of epoxy–glass composite*
(For solutions, see p. 256 and References 48–54)

The dynamic mechanical measurements on plastics are the subject of ASTM D4092 [48].

While the addition of any filler to a polymer will affect the chemical and mechanical properties of the material, the reinforcement of a polymer matrix with fibre material, such as glass fibre or carbon fibre, gives greatly increased strength. One glass fibre type frequently used is 'E-glass', which has a tensile modulus, $E \approx 100$ GPa, and a tensile strength around 2 GPa. It is not surprising that a polymer, such as nylon 66, which has $E \approx 1$–2 GPa and tensile strength around 70 MPa, is greatly increased in strength [49,50,51].

The polymer component may be a thermoplastic, such as nylon, or a thermoset such as a cross-linked polyester or an epoxy resin. The reinforcement can be of glass, carbon, mineral or metal fibres. The use of pre-impregnated materials or 'prepregs' makes the fabrication simpler, and often it is these materials which must be investigated for their cure kinetics. As the curing process proceeds, the glass transition temperature of the composite increases. The presence of glass fibre may affect the thermal stability of the material.

At temperatures below the glass transition the cured composite behaves like a glass. Above T_g, the composite is viscoelactic.

Sample:	epoxy–glass composite
Holder:	3-point bending platform
Rate:	2 K/min
Atmosphere:	nitrogen
Size:	23 mm × 3 mm × 0.9 mm
Frequency:	10 Hz
Static stress:	11 MPa

Questions

(a) What temperature is the glass transition? What other transitions may be observed?
(b) Given the values of the storage modulus E' and tan δ, calculate the loss modulus E'' at 100 and 150 °C.
(c) If this sample were only partially cured, what would happen to the moduli and to the temperature of the glass transition?

6.2.7   *Characterisation of a thin adhesive film*
(T.J. Lever, T.A. Instruments Ltd)
(For solutions, see p. 256 and References 55–57)

Coloured polyethylene terephthalate (PET) films are often coated on car windows to reduce heat build-up inside the car during exposure to strong sunlight. These films are typically 25 μm thick and are affixed to the glass using a thin (4 μm) layer of adhesive. Since the satisfactory performance and longevity of these films is obviously related to their ability to remain adhered to the glass, the glass transition temperature, T_g, of the adhesive is important, particularly where the windows will be exposed to harsh winter temperatures.

The characterisation of the adhesive sets a problem since it is a small component of the bulk material. It is not possible to find the T_g by DSC or TMA, or even DMA, since the effect is swamped by the thicker PET film. Even extracting the adhesive with a solvent was not suitable, since the solvent was an effective plasticiser!

Many adhesives contain polar groups – for example, epoxies, silicones or cyanoacrylates [55]. Since dielectric thermal analysis (DETA or DEA) detects relaxation processes at the molecular level, especially for polar groups, this technique should provide a higher sensitivity for this problem.

6.3
Fine chemicals and
pharmaceuticals

6.3.1   *Purity determination (ASTM E928)*
(For solutions, see p. 257 and References 58–61)

Background and method

As discussed in Chapter 3, the presence of impurities in a sample will cause the melting peak measured by DSC to broaden and to shift to lower temperatures. For pharmaceuticals, insecticides and fine chemicals, the assessment of purity is very important.

The ability to perform the experiment, together with the calculations, allows the analyst to check that the purity is within acceptable limits or to suggest the remedies needed.

In many cases, the purity measurement may be controlled by the computer attached to the DSC system but the corrections that must be made to the 'raw' data must be appreciated.

1. The apparatus is calibrated by running a suitable standard, such as high-purity indium (T_m = 156.6 °C, ΔH_{fus} = 28.7 J/g) under the same conditions as we shall use for the sample (for example, 0.5 K/min).
2. The sample pan is cleaned with a solvent (CH_2Cl_2) and dried at 700 K for 1 min. A sample of 1–3 mg is weighed into the pan, which should then be sealed to avoid mass loss. It has been found that placing an inner liner to keep the sample in good contact with the base of the pan

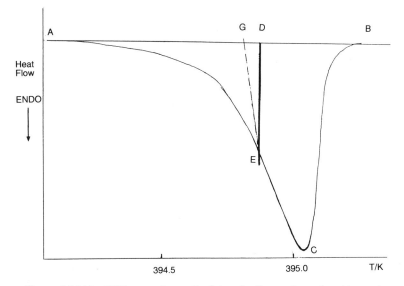

Figure 6.3.1(a) DSC curve for purity determination on benzoic acid sample.

improves reproducibility. The sample is then heated in flowing nitrogen through the melting transition at a low heating rate and the DSC curve recorded. A rate of 0.3–0.7 K/min is recommended, but higher rates are sometimes used. The pan is re-weighed to check that no loss has occurred.

3. The total area ABCA under the curve in Figure 6.3.1(a) is measured as well as the fractional areas such as ADEA for several cases in the range 10–50% of the total area.

4. The leading edge of the indium DSC peak has a definite slope. Each sample temperature recorded should be corrected by transposing a line of this slope through the point E to the baseline at G. At temperature G the fraction represented by the area ADEA has melted:

$$F = ADEA/ABCA$$

5. It is probably as well to check that the sample has not decomposed during the melting! If this is suspected, it is sometimes possible to try a short purity determination with the lower temperature points.

6. Corrections to the measured purity curve are estimated in order to produce a straight line plot of T versus $1/F$. This may be done by successsive additions, or by computer. The ASTM method advises working with fractions between 0.1 and 0.5 of the total area.

Problem

A sample of benzoic acid was thought be be slightly impure. A purity determination was carried out using the standard ASTM procedure, and

Table 6.3.1(a)

Corrected T/K	Fraction area/units	F	$1/F$
394.39	35.8	0.101	9.90
394.46	43.9	0.124	8.06
394.53	50.2	0.142	7.05
394.61	61.6	0.174	5.75
394.65	70.4	0.199	5.03
394.73	87.8	0.248	4.03
394.80	113.1	0.319	3.13
394.93	175.2	0.495	2.02

curves (Figure 6.3.1(a)) and results (Table 6.3.1(a)) were obtained.

Sample:	benzoic acid (C_6H_5COOH, RMM 122)
Crucible:	aluminium
Rate:	0.5 K/min
Atmosphere:	nitrogen, 20 cm^3/min
Mass:	2.40 mg
Total peak area:	354.0 units
Calibration constant:	0.904 mJ/unit

Questions

(a) Plot T against $1/F$ for the data given above. Attempt corrections by adding 10 or less to the fractional area *and to the total area* to linearise the graph.
(b) Calculate the corrected ΔH_{fus} per mole of benzoic acid. As a rough guide, enthalpies of fusion are of the order of 10–35 kJ/mol.
(c) Calculate the percentage purity of the sample.
(d) Suggest what methods might be used
   (i) to check the percentage purity and
   (ii) to identify the chemical nature of the impurities.

6.3.2 *Phase diagrams of organic chemicals*
(For solutions, see p. 260 and References 62–66)

The phase behaviour of mixtures of chemicals is very important in the fields of pharmaceuticals [62,63,64], fine organics [65] and in metallurgy [66]. In industrial situations, the interaction, solubility and crystallisation may govern the processing method to be used.

Urea ($CO(NH_2)_2$, RMM 60) and phenols are known to form compounds [62,64]. The problem here is to investigate the phase behaviour and to find any compound formation with *m*-nitrophenol (*m*-$NO_2 \cdot C_6H_4 \cdot OH$, RMM 139).

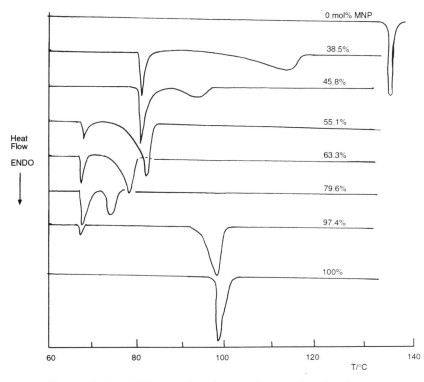

Figure 6.3.2(a)   DSC curves for mixtures of urea and *m*-nitrophenol.

Table 6.3.2(a)

Tube	U/g	MNP/g	Tube	U/g	MNP/g
1	0.500	0.000	5	0.100	0.400
2	0.194	0.285	6	0.050	0.450
3	0.144	0.282	7	0.003	0.248
4	0.100	0.285	8	0.000	0.500

Mixtures of urea and *m*-nitrophenol were made up by accurately weighing into small glass tubes amounts of each to give a total mass of about 0.5 g. The components were mixed thoroughly and then melted by gentle heating. After cooling, samples of about 4 mg were run on the DSC (Figure 6.3.2(a) and Table 6.3.2(a)).

Samples: mixtures of urea (U) and minitrophenol (MNP)
Crucible: aluminium
Rate: 8 K/min

Atmosphere: nitrogen, 10 cm^3/min
Mass: $\approx$ 4 mg

Questions

(a) Measure the extrapolated onset temperature, T_1 and the final temperatures of any subsequent peaks, T_2, etc.
(b) Plot the phase diagram and label as fully as possible.
(c) If a *third* component were added, would it still be possible to determine the phase diagram? What other thermal techniques could be used here?

6.3.3   *Liquid crystal studies*
(J.W. Brown, Kingston University)
(For solutions, see p. 260 and References 67–71)

Introduction

Liquid crystals possess a degree of molecular order intermediate between the three-dimensional order of a crystal and the complete disorder of a true liquid [67,68]. Liquid crystals may be 'thermotropic', that is, they are formed by heating certain crystalline materials such as *p*-azoxyanisole (see p. 189), or 'lyotropic', formed by treating certain compounds, such as soaps, with a suitable solvent.

Considerable interest has been shown in these materials recently because of their potential use in opto-electronic devices and also for flat-screen displays for computers and televisions [69]. New compounds are being synthesised and tested for liquid crystalline properties by many techniques, especially DSC and thermomicroscopy [70]. The effects of chemical structure on their transition temperatures and on the phase behaviour of pure compounds and mixtures must be investigated. The molecules are often elongated systems with a fairly rigid, polar structure – for example, a molecule with linked, *p*-substituted aromatic rings.

Three broad types of liquid crystalline phase are identified. The highest degree of order, after crystals, is shown by the *smectic phases*, which have a layered structure with molecules aligned at a common angle, parallel to each other within the layers. The many different polymorphic smectic phases may have different alignment angles and order within the layers, and show different optical, X-ray and thermal behaviour.

If the molecules have their long axes parallel, but are randomly arranged, we get the *nematic phase*, with a characteristic optical texture of lines and brushes. *Cholesteric phases* have a sequence of nematic layers with a progressive change in the direction of the axis in going from one layer to the next, 'like a pack of cards, twisted about an axis normal to the plane of the cards' [67].

Problem

A new material (**III**) was synthesised with *p*-di-substituted aromatic rings.

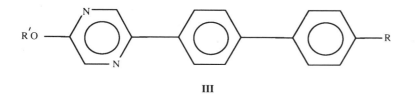

III

$$R' = C_9H_{19}O; \ R = C_3H_7$$

When run on the hot-stage polarising microscope it gave the photographs shown in Figure 6.3.3(a) from which the textures may be observed, and on the DSC the curves (Figure 6.3.3(b)).

Sample:	new material 'APPPP', crystalline
Crucible:	aluminium (DSC); glass (HSM)
Rate:	5 K/min
Atmosphere:	static air
Mass:	≈ 1 mg

Questions

(a) Does this material show any liquid crystalline behaviour?
(b) If so, what types of phases are present?
(c) What other techniques might be used to confirm the phases?

6.3.4   *Stability and polymorphism of pharmaceuticals*
(For solutions, see p. 261 and References 72–76)

The stability, and hence the shelf-life, of pharmaceutical drugs and of dosage mixtures is highly important since the product may become less effective or even produce by-products which are less beneficial or cause side-effects.

   In section 6.3.1 we examined a method of testing the purity of the sample. Clearly this could be applied to actual samples that have been stored for extended periods at a specified temperature (e.g. 80 °C) and this is often a legal requirement. The time taken for the assay to fall to 90% of its initial value is called $t_{0.1}$. This type of test takes a considerable time.

   In this section we shall try to set up suitable thermal analysis tests to aid in testing of pharmaceutical products.

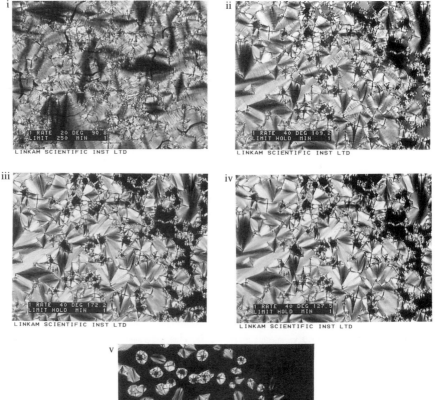

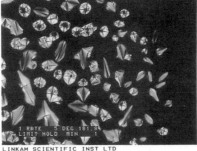

Figure 6.3.3(a)   Hot-stage microscope photographs of APPPP. Linkam hot-stage, crossed polars. (i) 90 °C; (ii) 109 °C; (iii) 127 °C; (iv) 172 °C; (v) 182 °C.

Questions

(a) How could we estimate the stability of a drug compound in a shorter time?

(b) How could we determine whether the sample degrades at melting, or has polymorphic forms?

(c) What methods could be used to study the loss of water or of solvent from the sample?

(d) How much sample shall we need for each of these tests?

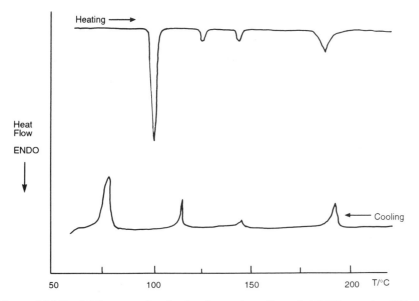

Figure 6.3.3(b)   DSC curves for the heating and cooling of APPPP sample. 5K/min, 1 mg, static air.

6.3.5   *Dynamic mechanical analysis of food products*
(Courtesy TA Instruments)
(For solutions, see p. 263 and References 77–81)

Samples of food materials may deteriorate unless they can be kept under 'ideal conditions'. The absorption of water, or its loss, will alter the texture, appearance and the appeal of the product. Samples of food with the wrong rheological or mechanical characteristics may be unsuitable for some purposes.

While such thermal processes as protein denaturation and starch gelatinisation may be studied by DSC [77,78,79] and loss of moisture could be followed by TG, the small changes that these cause in mechanical properties of the food are best studied by DMA [80,81].

Sample: Commercial white bread, exposed to atmosphere, 23 °C, 70% relative humidity; compressed from 14 mm thick to 2 mm
Clamping: vertical clamps; strip 8 mm × 13 mm × 2 mm
Rate: 2 K/min
Atmosphere: air

Questions

(a) What happens to the bread samples?
(b) What would be the next test you might do?

6.4
Other materials

6.4.1 *Carbon oxidation*
(D. Dollimore, Department of Chemistry and College of Pharmacy, University of Toledo, Toledo, Ohio, USA)
(For solutions, see p. 263 and References 82–85)

Introduction

Carbons may be prepared by heating certain organic compounds in the absence of oxygen. If the compounds have a low molecular mass, e.g. methane, then a carbon black results. If, on the other hand, the organic compound is a macromolecule, then the action of heat in the absence of oxygen may *either* break down the polymer by chain scission to produce monomers or fragments of lower molecular mass, *or* cause the polymer to degrade by chain stripping to produce a carbon skeleton. In the latter case, the carbon product may be produced and retain the solid state throughout to give a *char* or *charcoal*. In other instances, carbonisation is accompanied by a fluid transformation, and in this case the product is termed a *coke*.

Both cokes and charcoals may be 'activated' to produce a high-area activated carbon by thermal treatment in air, oxygen, carbon dioxide or water vapour. The activation process involves the gasification of a few percent of the material to develop a high-area, porous structure. Chemical activation with metal oxysalts produces extremely high surface area carbons.

The gasification of carbons in air occurs with the formation of carbon monoxide, carbon dioxide and traces of other gases due mainly to impurities. There is a correlation between the activity and surface area which may be investigated by following the loss of mass of the solid phase [82]. The correlation is seen most easily in carbon blacks where all the surface is an 'external surface' with no surface residing in pores.

A temperature jump method was used because of its simplicity. A sample is placed in the thermobalance and heated to the starting temperature. The temperature is held for a given time, is then increased by 10 K, and held constant for the same time interval as before. The process is repeated until six good plots of mass against time have been completed. The time interval (20 min in this case) was chosen carefully so that in this time a *linear* plot of mass against time was recorded, which greatly simplifies the evaluation of the Arrhenius parameters. Experimentally, we note that zero-order kinetics apply:

$$d\alpha/dt = k_T$$
$$\text{and } \alpha = k_T t$$

where α is the fraction decomposed, t is time and k_T is the specific reaction rate at temperature T K.

Experimental (Figures 6.4.1(a) and (b))

To complement the kinetic measurements, the surface area of each sample of carbon black was determined by nitrogen adsorption.

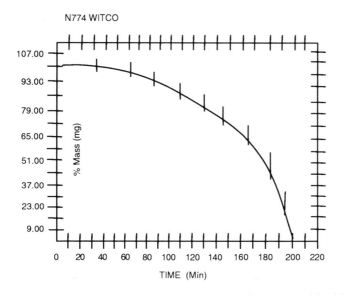

Figure 6.4.1(a)   Temperature jump plot for carbon black Witco N774 oxidised in air using 20-min periods at each temperature.

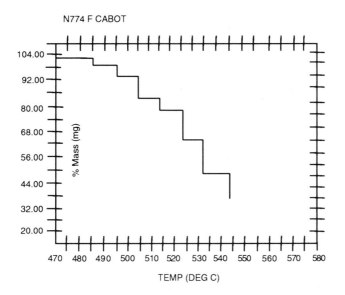

Figure 6.4.1(b)   A typical plot of mass loss vs temperature for carbon black Cabot N774 oxidised in air using the temperature jump method using 20-min periods at each temperature.

Table 6.4.1 Typical results of the oxygen jump method applied to carbon black Witco N683 of surface area 36 m^2/g

Mass loss %	T/K	$10^3 \ k/s^{-1}$
4.13	763	3.44
5.74	773	4.78
8.4	783	7.00
11.62	793	9.68
15.61	803	13.0
20.02	813	16.7

Samples: carbon blacks, dried for 1 h at 100 °C
Crucible: platinum
Rate: combined: isothermal steps and rising temperature
Atmosphere: air
Mass: 10 mg

Questions

(a) Figure 6.4.1(a) appears as a curve! Explain why this is so and how the rate constant could be calculated from the plot.
(b) From the data of Table 6.4.1, construct an Arrhenius plot and determine the activation energy for the carbon black Witco N683
(c) If a series of carbon blacks from different sources and with different surface areas were run, how could the results be correlated?

6.4.2 *Proximate analysis of coals*
(For solutions, see p. 266 and References 90–93)

The composition of a complex natural material like coal was formerly done by a series of tests to determine the moisture content, the volatile matter, fixed carbon and ash contents. These establish the grade or rank of the coal and are reported as the 'proximate analysis' (Figure 6.4.2(a)). The ASTM method D3172 reports the procedure for performing these measurements, but it has been shown that thermogravimetry can give a good proximate analysis in rather less time [90,91].

The procedure is to measure the thermogravimetric curve while heating to 110 °C in nitrogen for 5 min and then heating rapidly to 900 °C in nitrogen. After holding for a few minutes in nitrogen at 900 °C, the gas is switched to oxygen.

Sample: anthracite coal, coarse powder
Crucible: platinum
Rate: as above; 50 K/min
Atmosphere: nitrogen, 50 cm^3/min, then oxygen at 50 cm^3/min
Mass: 25.0 mg

Questions

(a) Measure the moisture content by the mass loss at 110 °C.
(b) Measure the volatile content by the mass loss between 110 and 900 °C.
(c) Measure the fixed carbon by the mass loss at 900 °C after switching to oxygen.
(d) Measure the ash content as the residue at 900 °C.

6.4.3 Oil testing
(For solutions, see p. 268 and References 94–100)

Introduction

The thermal and oxidative stability of mineral oils affects their performance and lifetime as lubricants and, in a similar way, the thermal and oxidative stability of edible oils affects their performance, taste and interaction with other foods. In either case, anti-oxidants may be added to prolong the lifetime of the oil.

In addition to the oxidative stability, thermal methods may be used on mineral oils to look at the wax content and its dissolution, the effectiveness of any additives [94,95] and could also be used to investigate the volatility of low-boiling oils. Analysis of oil shales has been carried out by DTA and TG [96,97].

The oxidative stability of edible oils and fats may be treated in a similar way. Burros [98] has reviewed the applications of thermal analysis in food chemistry, and Hassel has compared the 'active oxygen method' (AOM) with TG and DSC techniques [99].

Pressure DSC will have the advantage of allowing operation under higher pressures of oxidant, while restricting the volatilisation of the material. If no PDSC is available, it may be more suitable to run in pure oxygen at 1 atmosphere pressure.

Problem

Samples of mineral oil were provided and their comparative oxidative stabilities required for comparison with a micro-oxidation test which also investigated deposit formation during the oxidation [100].

The effect of contact with metals, such as carbon steel, was important.

Samples:	lubricating oils
Crucible:	aluminium
Rate:	8 K/min to 400 °C
Atmosphere:	oxygen, 25 cm³/min
Mass:	about 2 mg

Method To investigate the oil oxidation and simultaneously the deposit formation or discoloration, the simultaneous DSC–reflected light intensity apparatus described in Chapter 5 was used (Figure 5.2.5, Figure 6.4.3).

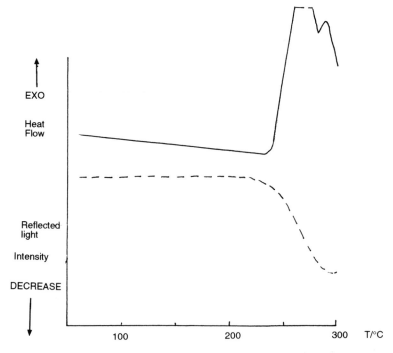

Figure 6.4.3 DSC and RLI traces for oxidation of an oil sample.

Note A problem arose immediately, since the oil became much less viscous as the temperature was raised and flowed into the edges of the crucible. This was solved by containing the oil on a small mat of glass fibre punched from a GF filter paper.

Questions

(a) Does the method detect onset of oxidation and of deposit formation, and are they different?

(b) What other techniques should be used to investigate this?

6.4.4   Soil analysis
   (For solutions, see p. 268 and References 101–103)

Soil samples must be analysed for their components, and classical methods require several independent tests, taking a considerable time. It has been claimed that analysis for moisture, organic content and mineral content may be carried out by thermogravimetry (Figure 6.4.4).

Samples: soils collected from forest area of Surrey, England, sieved through a 2 mm sieve and kept in sealed containers until used

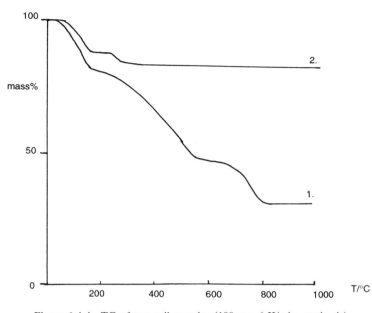

Figure 6.4.4 TG of two soil samples (100 mg, 6 K/min, static air).

Crucible: alumina
Rate: 6 K/min
Atmosphere: static air
Mass: about 100 mg

Questions

(a) Besides thermal analysis, what other tests should be performed on the original sample?
(b) Estimate the moisture content, organic matter and mineral residue of these samples.
(c) Is it possible to identify the minerals present? What other methods could be used, and what problems would occur if the soil had a high clay content?

6.4.5 *Catalyst studies*
(For solutions, see p. 269 and References 104–109)

Catalyst preparation by heating a suitable precursor in an active atmosphere was mentioned in Chapter 2. A DTA apparatus adapted to allow gas flow through a sample has been used to investigate the effectiveness of V_2O_5 catalysts for the oxidation of SO_2 [104] and a combined DSC–MS system using the Setaram Calvet microcalorimeter was employed to study the catalytic reduction of NO with NH_3 [105]. A flow-through TG system could be used with open or gauze crucibles.

The effects of the catalyst surface are very important, and measurements of the surface area paralleling the thermal and compositional analysis should be considered.

Adsorption of the starting materials onto the catalyst surface and desorption of the products may be investigated by thermal methods [106]. Chemisorption is generally an exothermic process, and desorption endothermic.

For the methanation of carbon monoxide:

$$CO + 3H_2 = CH_4 + H_2O$$

a nickel/alumina catalyst is used [107]. The performance depends on the nickel particle size and the mechanism of reaction at lower temperatures could be hydrogenation of adsorbed CO, followed by dehydration, or, at higher temperatures, CO dissociation followed by hydrogenation of the surface carbon. The reaction is highly exothermic ($\Delta H = -206$ kJ/mol) and could be investigated by DSC.

Samples: reduced $Ni–Al_2O_3$ catalyst
Crucible: aluminium
Rate: 10 K/min
Atmosphere: H_2, with CO added 40 cm³/min
Mass: 1–10 mg

Questions

(a) Discuss the shape of the DSC curves (Figure 6.4.5).
(b) How would the metal of the DSC cell affect the results?
(c) How could the kinetics of the reaction be studied?

6.4.6 *Survival of Antarctic micro-arthropods*
(W. Block and M.R. Worland, British Antarctic Survey, Cambridge)
(For solutions, see p. 269 and References 110–112)

The Antarctic environment presents the local population of mites and collembolans with a severe survival problem. Freezing temperatures and drying conditions may not occur together [110]. They survive by using two processes to improve their cold-hardiness. Firstly, they evacuate food and water from their guts to reduce the probability of heterogeneous nucleation. Secondly, they accumulate polyhydric alcohols such as glycerol to enhance their supercooling [110, 110a]. Antifreeze proteins inhibit the growth of ice below the equilibrium freezing point, and give a differential between the melting and freezing points [110a].

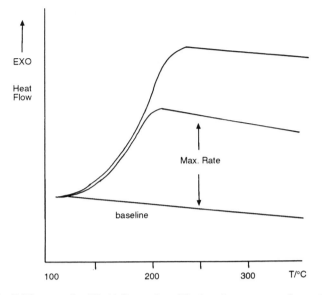

Figure 6.4.5 DSC curves for Ni–Al$_2$O$_3$ catalyst. The baseline was run after reduction in 100% hydrogen. The reaction curves were then obtained in (a) 0.2% and (b) 0.5% CO in hydrogen.

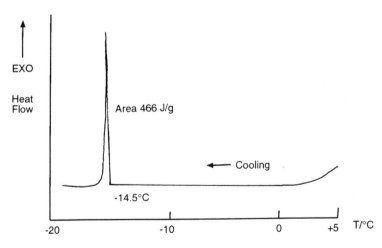

Figure 6.4.6 DSC curve for *Cryptopygus antarcticus* cooled at 1 K/min. (Data courtesy British Antarctic Survey and TA Instruments Ltd.)

The water loss could be studied by a modified isothermal TG experiment, where the species are maintained at constant temperatures in the range −10 to 45 °C over a desiccant and their mass loss measured [110]. For example, the species *Cryptopygus antarcticus* lost 5% water per hour at −10 °C (Figure 6.4.6).

The freezing studies could also be conducted using DSC [110a–112]. Using a cooling rate of 1 K/min, a freezing-intolerant mite froze about 80% of its body water, whereas a freezing-tolerant midge only froze 57% of its water.

Sample: whole insect *Cryptopygus antarcticus*
Crucible: sealed large aluminium
Rate: 1 K/min *cooling*
Atmosphere: He, 30 cm^3/min
Mass: 0.06 mg

Questions

Discuss the observations of the DSC curve (Figure 6.4.6).

Solutions

6.1.1   Tin(II) formate decomposition

(a) The TG shows clearly that there are two main stages of reaction:
   (i)   a loss of about 35% near 200 °C and
   (ii) a gain of mass starting near 600 °C.
   These could represent the decomposition of the formate to oxide:

$$Sn(HCOO)_2 = SnO + HCHO + CO_2$$

which gives a calculated loss of 35.5%, followed by the disproportion-ation and oxidation of the tin(II) oxide:

$$2SnO = SnO_2 + Sn \qquad\qquad (I)$$
$$2SnO + O_2 = 2SnO_2 \qquad\qquad (II)$$
$$Sn + O_2 = SnO_2 \qquad\qquad (III)$$

A residue of tin(IV) oxide would correspond to a final mass loss of around 28%, but in practice it is about 29–32%. It is not possible to detect the disproportionation reaction by TG.

   DTA confirms these changes, giving an endothermic (double) reaction near 200 °C and a broad exotherm near 600 °C.

   The value of ΔH for the disproportionation reaction (I) calculated at 298 K is −9.1 kJ/mol, and the oxidation reactions are both very exothermic.

(b) The double DTA peak near 200 °C corresponds to a change in slope of the TG, but no definite intermediate could be detected by X-rays. However, examination in a hot-stage microscope showed that the sample *melted* during the first endotherm. This is the fusion of the original salt, which then decomposes more rapidly in the molten state. Unusually, at a slower heating rate, a single peak is obtained because the salt may decompose completely before it can melt.

(c) The products of reaction have been confirmed by several methods:
   (i)   The gaseous products released around 200 °C were studied by mass spectrometry and shown to be HCHO, HCOOH (by oxidation of the methanal) and CO_2.
   (ii) The solid product was studied by scanning electron microscopy (SEM). It was seen that metallic tin formed first on the surface of the tin(II) oxide matrix and was then oxidised to tin(IV) oxide.
   (iii) X-ray analysis showed that only the formate and oxide were produced in the first stage and only tin metal, tin(II) oxide and tin(IV) oxide in the second stage.

6.1.2 *Mixtures of carbonates*

(a) As reported in Chapter 2, calcium carbonate decomposes above 600 °C, and so the high-temperature loss of 2.2% must be due to the $CaCO_3$. This corresponds to the reaction:

$$CaCO_3 = CaO + CO_2 \quad \text{Calculated loss: 44\%}$$

Thus a loss of 2.2% corresponds to $2.2 \times 100/44 = 5\%$ of the mixture.

(b) The amount of copper carbonate must therefore be 95% of the original mixture. The loss at the lower temperature of about 200 °C was about 27%. The calculated losses from the possible 'carbonate' formulae are:

$$CuCO_3 \qquad\qquad = CuO + CO_2 \qquad\qquad \text{Calculated loss: 35.6\%}$$
$$CuCO_3 \cdot Cu(OH)_2 \;= 2CuO + CO_2 + H_2O \quad \text{Calculated loss: 28.0\%}$$
$$2CuCO_3 \cdot Cu(OH)_2 = 3CuO + 2CO_2 + H_2O \quad \text{Calculated loss: 30.8\%}$$

The closest agreement is for $CuCO_3 \cdot Cu(OH)_2$. This is generally reported as the usual copper carbonate [2] although it may be hydrated. X-ray data confirm that this is the solid present.

Note that the DTG curve shows only a single loss, so both carbon dioxide *and* water vapour are lost simultaneously [4]

(c) Silver carbonate has been reported to decompose in two stages:

$$Ag_2CO_3 = Ag_2O + CO_2 = 2Ag + \tfrac{1}{2}O_2 + CO_2$$

The temperatures of these decompositions are about 230 and 400 °C, so that the first reaction would almost coincide with the copper carbonate decomposition, but the second would be 'in the clear'. The decompositions might be expected to be simple, but have been shown to depend upon the mode of preparation, and the conditions used [3]. Although a reasonably good calculation of the amounts of silver, copper and calcium carbonates can be made from the TG curve, the results are rather variable.

6.1.3 *Strontium nitrate decomposition*

(a) The TG run shows a single mass loss of 51%, which suggests that the nitrate goes eventually to the oxide:

$$Sr(NO_3)_2 = SrO + (NO, NO_2, O_2)$$

These three gases are all detected by the mass spectrometer:

nitric oxide,   NO, $m/z = 30$
nitrogen dioxide, NO_2, $m/z = 46$
oxygen, O_2, $m/z = 32$

(b) Since the oxygen is evolved at a lower temperature than the other gases, showing the very high sensitivity of the MS system, it is possible that there is a tendency to form the nitrite:

$$Sr(NO_3)_2 = Sr(NO_2)_2 + O_2$$

However, the evidence is inconclusive and there is no plateau or even a dip in the TG curve. The evolution of NO follows soon after the production of oxygen, so that the decomposition is continuous.

(c) The dip in the MS curves for NO and O_2 at around 590 °C was found by observation of the reaction on a hot-stage microscope to correspond to the bulk fusion of the partially decomposed nitrate [7]. This will affect the kinetics of decomposition. DTA studies show that strontium nitrate melts around 600 °C and decomposes endothermically with considerable bubbling with a second DTA peak at 730 °C [8].

Note 'The rate of decomposition was found to be strongly dependent on the thickness of the molten film and the efficiency of product gas removal. This extreme dependence upon experimental conditions has generated confusion in the literature regarding the mechanism of reaction. The use of multiple simultaneous techniques is proving extremely helpful in the resolution of these anomalies' [5].

6.1.4 *Calorimetry and phase transitions of potassium nitrate*

(a) The deflections must be corrected for the 'empty pan' baseline, so that

$$\Delta y(\text{sample}) = (y(\text{sample}) - y(\text{empty}))$$

The calculation of the heat capacity at temperature T for samples run *under the same conditions as calibrants, especially the same heating rate*, may be done using the formula:

$$C_p(\text{unknown}, T) = \frac{\Delta y(\text{unknown}, T) \times C_p(\text{calibrant}, T) \times m(\text{calibrant})}{\Delta y(\text{calibrant}, T) \times m(\text{unknown})}$$

Some values at specific temperatures are tabulated in Table 6.1.4. The method used is specified in ASTM E968.

Table 6.1.4

T (K)	C_p (J/(K g))		C_p(J/(K mol))
	Sapphire	KNO_3	KNO_3
350	0.873	0.973	98.4
370	0.904	1.018	102.9
380	0.918	1.039	105.1
390	0.931	1.060	107.2
410	0.955	TRANSITION	
440	0.987	1.240	125.4
450	0.997	1.245	125.8
470	1.015	1.258	127.2

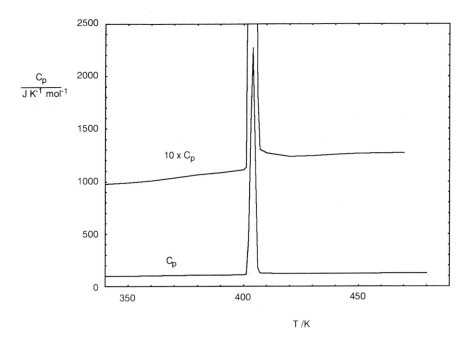

Figure 6.1.4(b) Heat capacity curves for potassium nitrate.

These are in reasonable agreement ($\pm$1%) with the results of Carling [10] over this range and are plotted out in Figure 6.1.4(b).

(b) Given the data for ΔH and T for the transition,

$$\Delta S_t = \Delta H_t / T_t$$
$$= 5350/401$$
$$= 13.3 \text{ J/(K mol)}$$

Literature values range from 12.4 [11] to 13.8 [10]. On average, we may write the *molar* heat capacities *over this range of temperatures* for KNO_3 (RMM 101.1) as:

$$C_p \text{ (Phase II)} = 21.4 + 0.220 \times (T/K) \text{ J/(K mol)}$$
$$C_p \text{ (Phase I)} = 99.0 + 0.060 \times (T/K) \text{ J/(K mol)}$$

and ΔH (II–I) = 5.35 kJ/mol at 401 K Using Kirchhoff's equation, $(\partial \Delta H / \partial T)_p = \Delta C_p$ and integrating, we get:

$$\Delta H\ (T_2) = \Delta H\ (T_1) + \int \Delta C_p\ dT$$
$$= \Delta H\ (T_1) + \Delta a(T_2 - T_1) + \Delta b/2 \times (T_2^2 - T_1^2)$$

From the above:

$$\Delta a = \ (99.0 - 21.4) \ = \ \ 77.6 \text{ J/(K mol)}$$

$$\Delta b = (0.060 - 0.220) = -0.160 \ J/(K \ mol)$$

Thus, for 298 K:

$$\Delta H(II-I, 298 \ K) = 5350 + 77.6 \times (298 - 401) - 0.080 \times$$
$$(298^2 - 401^2)$$
$$= +3117 \ J/mol = + 3.12 \ kJ/mol$$

The probable error in this value is about 1%. Therefore,

$$\Delta H_f \ (I, 298 \ K) = -497.1 + 3.1 = -494.0 \ kJ/mol$$

(c) (i) Using Hess's Law,

$$\Delta H \ (III-II) = \Delta H \ (III-I) + \Delta H \ (I-II)$$
$$= 2.66 + (-5.35)$$
$$= -2.69 \ kJ/mol$$

(ii) Again, using Hess's law:

$$\Delta H \ (II-l) \quad = \Delta H \ (II-I) + \Delta H \ (I-l)$$
$$= 5.35 + 11.0$$
$$= 16.35 \ kJ/mol$$

Note! Errors will certainly be present since we have not measured C_p down to the required temperature. Also, it should be noted that strict calorimetry using DSC really requires much more sophisticated calculations [13].

6.1.5 *Decomposition of barium perchlorate*

(a) (i) The loss of hydrate water takes place in two stages.

$$Ba(ClO_4)_2 \cdot 3H_2O \ (solid) = Ba(ClO_4)_2 \cdot 2H_2O + H_2O$$
$$Ba(ClO_4)_2 \cdot 2H_2O = Ba(ClO_4)_2 + 2H_2O$$

The total RMM = 390.4, so that the first reaction should show a loss of $18 \times 100/390 = 4.6\%$, and the second a total loss of 3×4.6 = 13.8% of the original mass. This behaviour is echoed by the DTA curve which shows two peaks, the second being a 'doublet', which might indicate three different environments for the three water molecules.

(ii) The DTA curve alone shows two sharp endotherms between 250 and 400 °C. These have been reported [16] as phase transitions of anhydrous barium perchlorate and have been used as a 'fingerprint' for the presence of barium perchlorate.

(iii) Both DTA and TG show a broad, double peak for the exothermic decomposition, in some cases preceded by a small endotherm and corresponding to a total mass loss of 46.6%

There is some evidence of a discontinuity in the traces, but this was found to disappear when a multiplate sample holder was used allowing free loss of product gases.

The product could be barium oxide or barium chloride:

$$Ba(ClO_4)_2 \cdot 3H_2O = BaO + Cl_2 + 3.5O_2 + 3H_2O \text{ Total loss } 60.7\%$$
$$Ba(ClO_4)_2 \cdot 3H_2O = BaCl_2 + 4O_2 + 3H_2O \qquad \text{Total loss } 46.6\%$$

Obviously, the chloride is the more likely product.

(b) The loss of water could be confirmed by specific water detectors (see Chapter 5), by IR or MS. The phase changes and the final barium chloride are confirmed by X-ray studies.

(c) The effect of MnO_2 and other semiconducting oxides is catalytic. The water is lost in a different manner, suggesting different environments. The first stage of the final decomposition was found by kinetic analysis [15] to have an activation energy of 110 kJ/mol in the presence of MnO_2. When run alone E_a was 225 kJ/mol. The mechanism of reaction was also changed.

6.1.6   Solid-state reactions

(a) The mass losses around 100 °C may be attributed to irreversible loss of moisture from the reagents. They correspond to the endotherms 1 and 3 on the DTA trace. The only *exothermic* peak, Peak 4, might be due to a solid-state reaction:

$$BaCl_2 + 2KNO_3 = Ba(NO_3)_2 + 2KCl$$

This is not reversible, and consequently does not appear in re-runs. Calculation of ΔH for the reaction from published data [22] gives a value of about -20 kJ/mol corrected to 270 °C.

The large, sharp, endothermic Peak 5 does appear in the re-run, and must be due to the fusion of the mixture of products at 320 °C.

The products are confirmed by X-ray diffraction studies, and also by running a 1 : 2 mixture of $Ba(NO_3)_2$ and KCl, which only shows Peak 5.

Since the reactions are (i) at high temperature and (ii) not reversible, they would perhaps not be suitable for solar energy storage. Suitable systems are discussed in [23].

(b) Grinding the reactants together improves the surface contact, reduces the particle size, and 'mechanically activates' the solid-state reaction. Fatu [21] has made measurements of the enthalpy and of the activation energy of the reaction.

(c) Peak 2 corresponds to the solid–solid phase transition of KNO_3, at 128 °C, which was discussed in Problem 6.1.4. Since the potassium nitrate disappears in the reaction, this peak too will disappear.

6.2.1 *Characterisation of a polymer*

(a) While the scale of the figures does not allow highly accurate measurements, the temperatures during the melting and crystallisation peaks and the glass transition are approximately:

Melting/°C		*Crystallisation/°C*		*Glass transition/°C*	
T_f	185	T_f	163	T_f	50
T_m	200	T_c	143	T_e	63
T_e	210	T_e	131	T_m	58

Identification using a suitable transition range eliminates many polymers, but gives as possibilities *nylon 6* or *nylon 6/10* which have melting ranges around 200 °C and T_g values around 50 °C.

Measurement of the heat capacity might give additional confirmation, as might the study of the crystallisation behaviour [27,28].

(b) The TG trace shows a mass loss of about 5% at low temperature, which is probably moisture. This suggests a polymer having polar groups, such as a nylon. The decomposition temperature for 1% loss is around 350 °C, which is in accord with reports for nylon 6 [29].

Note   DMA experiments on this material gave a storage modulus value of about 1 GPa at 25 °C , a tan δ peak around 50 °C for the glass transition and a broad beta transition around −50 °C.

Conclusion   The most probable identification of this material is *nylon 6* without plasticiser or filler. It does not show any multiple melting peaks in this case. Further analysis (for example, by IR) would be most useful.

6.2.2 *Polymer blend analysis*

(a) Sample 1 has a majority of HDPE, melting at around 125 °C and the small peak at 160 °C is due to the PP impurity. The peak areas given show that:

$$\text{for}\quad \text{HDPE} = 1663/10.5 = 158.4 \text{ J/(g mixture)} \quad \text{and}$$
$$\text{for}\quad \text{PP}\quad = 126/10.5 \;= 12.0 \text{ J/(g mixture)}.$$

Since *pure* PP has a ΔH_m of 100 J/g, this sample must have a PP content of $100 \times 12.0/100\%$ PP = 12.0% PP by weight.

Note: Calculating on the PE ($\Delta H_m = 180$ J/g) we find a PE content of $100 \times 158.4/180 = 88\%$ by weight.

Sample 2 is mainly PET, and does show the glass transition peak around 80 °C as well as the main melting at 250 °C. The minor peak at 170 °C is PP, and the percentage is calculated as above:

For PP: Peak = 154/17.1 = 9.0 J/(g mixture) and so
PP = 100 × 9.0/100 = 9.0% PP
For PET: Peak = 622/17.1 = 36.4 J/(g mixture) and so
PET = 100 × 36.4/40 = 91% PET.

For a series of samples, it is probably better to establish a calibration graph of $\%A$ versus peak area.

The mechanical properties of the samples are of prime importance in their recycling and re-use. Measurements of the DMA curves and comparison with standard material would be useful.

The presence of additives in the recycled material possibly from the 'impurity' polymer such as PP may have a deleterious effect on the moulding characteristics because it affects the crystallisation rate. This rate may also be studied by DSC.

(b) Analysis by other thermal methods might be possible, but complementary techniques, such as IR, might be more quantitative.

6.2.3   *Kinetic studies of polymer cure*

(a) Figure 6.2.3(c) shows the plots obtained for the different values of m and n. Reasonably good agreement is obtained for $m = 1$, $n = 2$ over the range of $\alpha = 0.05–0.40$. Evaluating the constants from this plot gives $K_1 = 0.42 \times 10^{-3}$ s^{-1} and $K_2 = 2.28 \times 10^{-3}$ s^{-1}.

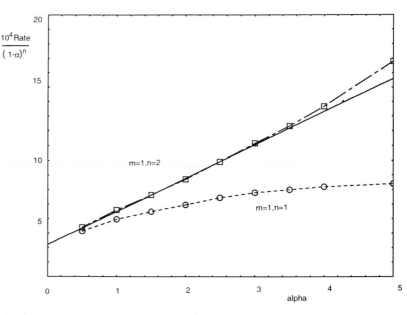

Figure 6.2.3(c)   Plots of rate function against alpha.

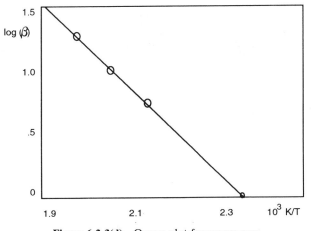

Figure 6.2.3(d)   Ozawa plot for epoxy cure.

A good test of the validity of the equation used is to plot a 'reduced time' graph of α against $t/t_{0.5}$ for several temperatures ($t_{0.5} = t$ where $\alpha = 0.5$). For this reaction there is good fit over the temperature range 150–200 °C.

Once the kinetics are established, it is possible to analyse an isothermal or a scanning DSC trace more fully.

(b) The plot of $\log_{10} \beta$ against $1/T_{peak}$ is shown in Figure 6.2.3(d). The slope is -3850 K. Therefore, $E = -R \times$ slope/0.4567 giving a value of $E = 70.1$ kJ/mol. This may be refined by using the Flynn and Wall method to give a better value of about 66 kJ/mol.

Note   ASTM method D5028 uses similar procedures to study the cure properties of pultrusion resins by thermal analysis.

6.2.4   *Polymer decomposition studies*

(a) The TG curve shows a small weight loss of 1.3% around 100 °C which is due to water evolved [42,43] during the initial reaction being trapped in the novolak. This gaseous product is confirmed by the $m/z = 18$ trace on the EGA. There is also a melting or glass transition reported below 100 °C [40,41]. These two events may be differentiated by running the DTA in sealed or high pressure cells to restrain the loss of water.

The major sharp mass loss around 140 °C accompanied by an exotherm on the DTA and a large peak at $m/z = 17$ on the EGA corresponds to the curing reaction shown in Scheme 6.1.

The accompanying loss of water and of a little formaldehyde

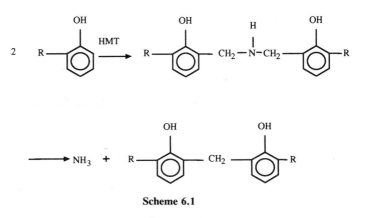

Scheme 6.1

($m/z = 29$) could be due to release of further trapped or lightly bound material.

Degradation, accompanied by further cross-linking, goes on in stages from about 250 °C. The release of more ammonia, and later of water and formaldehyde, is complementary to the mass loss shown by TG and the broad DTA peaks.

(b) The curing and kinetics have been studied [43] by using isothermal TG at various temperatures between 140 and 240 °C and by scanning TG. The release of ammonia could also be monitored by EGA, either mass spectrometric [41] or chromatographic [40].

(c) A fully finished product would not be expected to show the initial water loss, or the sharp curing reaction. However it would show the later, post-cure stages, and it has been found that release of water, ammonia and formaldehyde causes cracking of the sample and spikes on DTA and DTG traces due to very rapid losses. At temperatures above 400 °C, phenol ($m/z = 94$) was detected in addition to other products.

6.2.5 Oxidative stability of polymers

(a) The temperature scanning traces (Figure 6.2.5(a)) show that the outer sample (i) has a much lower induction temperature, T_{ind}, of 211 °C compared with the inner sample (ii) which has not been exposed to UV and has a $T_{ind} = 245$ °C. The indium melting peak occurs at the correct temperature. Specifications for piping set a satisfactory value of T_{ind} at 220 °C.

The same behaviour is found for the isothermal traces (Figure 6.2.5(b)), where the outer sample (iii) starts to oxidise after 15 min, but the inner (iv) does not do so until 55 min. The times may vary, but 'good samples' generally show an OIT over 30 min.

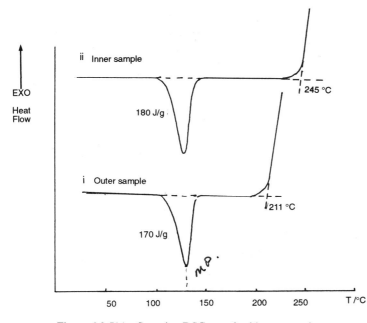

Figure 6.2.5(a) Scanning DSC runs for blue water pipe.

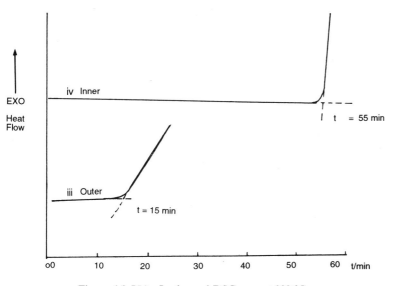

Figure 6.2.5(b) Isothermal DSC runs at 200 °C.

(b) Other information may be obtained from the DSC scan such as the melting peak of the polyethylene. The degree of crystallinity is found from the enthalpy of melting. For 100% crystalline material, $\Delta H_{fus} = 290$ J/g. Since we have values here of 170 J/g for (i) and 180 J/g for (ii), the outer, degraded sample is slightly less crystalline. This is discussed further in ASTM D3417.

The other tests that could be used to quantify the degradation of this polymer are many. Thermal methods such as TG, TMA and DMA will give additional information. Spectroscopic techniques such as IR to look at the carbonyl peak near 1700 cm^{-1} and UV to examine the formation of any conjugated structures have been used [46].

Notes   1. The catalytic effects of copper and copper salts require that polyethylene in contact with copper (e.g. in electrical wiring) should contain stabilisers to protect against copper. If such stabilisers are not present, OIT values as low as 0.5 min at 190 °C are reported [45].

2. In the recycling of polymers, the oxidative stability is important. The onset temperatures of blended recycled polymers showed that a fresh sample gave an onset at 291 °C, but with 45% recycled content, the onset was 293.7 °C. The rate of oxidation increased with recycled content [47].

6.2.6   *Studies of epoxy–glass composite*

(a) The glass transition of this material occurs between 120 and 175 °C (Figure 6.2.6). If we accept the *mid-point* of the major change as the T_g [48], then we would estimate it as 150 ± 5 °C.

Below the T_g we observe another peak, more clearly in the tan δ curve, due to the beta transitions at about −20 °C [52,53].

(b) Since tan δ = E''/E', we may calculate:

At 100 °C: $E' = 12.3$ GPa,   tan δ ≈ 0.018,
so $E'' ≈ 0.22$ GPa.
At 150 °C: $E' = 3.1$ GPa,   tan δ ≈ 0.33,
so $E'' ≈ 1.02$ GPa.

(c) As the sample is cured, the modulus and the T_g both increase. Isothermal curing shows an increase in modulus with time, and increasing the post-cure time increases T_g.

6.2.7   *Characterisation of a thin adhesive film*

The film sample was placed between the parallel plates of the DEA apparatus and under slight load to maintain good contact. The samples were heated from −100 to +50 °C and the multiplexed DEA curves

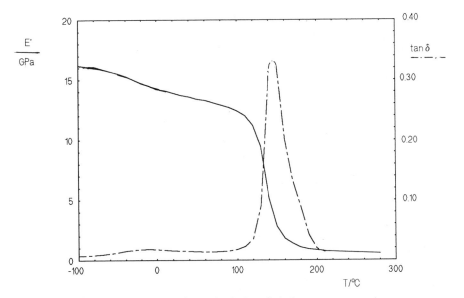

Figure 6.2.6 Dynamic mechanical analysis for epoxy composite.

recorded at frequencies between 1 Hz and 1 kHz. The results are shown in Figure 6.2.7.

Figures 6.2.7 (a) and (b) show the dielectric loss factor ϵ'' as a function of temperature for the polyester with and without adhesive coating. The PET substrate showed a broad, frequency-dependent transition, designated as a beta transition with a peak between about −90 and −20 °C. The PET adhesive sample, on the other hand, exhibited the PET beta transition, but also showed a higher temperature transition at about −20 °C at 1 Hz, which is the T_g of the adhesive.

Figure 6.2.7(c) presents the dielectric permittivity ϵ' of the coated film which shows an inflection point at the T_g of the adhesive which is not seen in the trace for PET alone. Either of these parameters could be used to characterise the adhesive.

6.3.1   Purity determination (ASTM E928)

(a) The plot of T against $1/F$ is not linear, so must be corrected. Normally, we should use computer software provided by the instrument manufacturer, but here we wish to check the method manually.

Attempts to straighten the line by adding successive increments equivalent to a fraction of 0.025 should allow a reasonable number of calculations. In this case 10 units is 0.028 of the total area. The graph (Figure 6.3.1(b)) and Table 6.3.1(b) show the results of successive additions of 10 units or less.

6.3
Fine chemicals and
pharmaceuticals

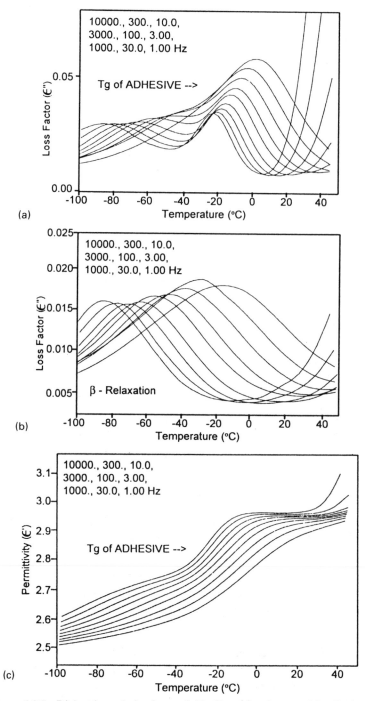

Figure 6.2.7   Dielectric analysis of coated thin film: (a) polyester with adhesive, ϵ'';
(c) polyester with adhesive, ϵ'.

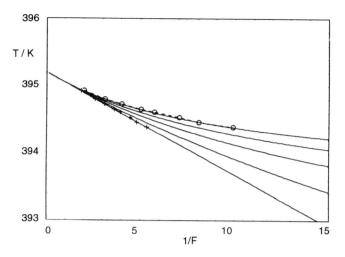

Figure 6.3.1(b)   Plots of $1/F$ (corrected) against T.

Table 6.3.1(b)   Correction of $1/F$

Addition/units:

0	10	20	30	35	37.5
9.90	7.95	6.70	5.84	5.49	5.34
8.06	6.75	5.85	5.20	4.93	4.81
7.05	6.05	5.33	4.79	4.57	4.46
5.75	5.08	4.58	4.19	4.03	3.95
5.03	4.53	4.14	3.82	3.69	3.63
4.03	3.72	3.47	3.26	3.17	3.12
3.13	2.96	2.81	2.68	2.63	2.60
2.02	1.97	1.92	1.87	1.85	1.84

The plots become more linear, and that with an addition of 37.5 units (fraction 0.106) is a good straight line.

$$\text{Slope} = -0.1530 \text{ K}$$
$$\text{Intercept} = 395.2 \text{ K} = T_0$$

(b) The correction of 0.106 gives a corrected total area of:

$$\text{Area} = 1.106 \times 354.0 = 391.5 \text{ units}$$

and thus

$$\begin{aligned}\Delta H_{\text{fus}} &= \text{Sensitivity} \times \text{Area} \times \text{RMM/mass} \\ &= (0.904 \times 103 \times 391.5 \times 122)/(2.40 \times 10^{-3}) \\ &= 17\ 995 \text{ J/mol}\end{aligned}$$

(c) Since the slope is $-(RT_0^2/\Delta H_{\text{fus}})x_{\text{B}}$, where x_{B} is the mole fraction impurity, then:

Table 6.3.2(b)

Tube	U/g	MNP/g	mol% MNP	$T_1/°C$	$T_2/°C$
1	0.500	0.000	0.0	133	–
2	0.194	0.285	38.5	79	112
3	0.144	0.282	45.8	79	91
4	0.100	0.285	55.1	67	80
5	0.100	0.400	63.3	67	77
6	0.050	0.450	79.6	66	74
7	0.003	0.248	97.4	65	94
8	0.000	0.500	100.0	96	–

$$x_B = 0.1530/(8.314 \times (395.2)^2/17\,995)$$
$$= 0.00212 \text{ or}$$
$$\text{Purity} = 99.79\,\%$$

(d) The purity could be checked by chromatographic or spectroscopic methods which could also identify the impurity after suitable calibration.

6.3.2 *Phase diagrams of organic chemicals*

(a) The temperatures measured from the curves with an accuracy of ± 1 °C and assumed to be corrected for any thermal lag are given in Table 6.3.2(b).

(b) The phase diagram (Figure 6.3.2(b)) is typical of systems forming a compound with an *incongruent melting or peritectic point*. From the limited number of runs reported here, it seems probable that the components form a 1 : 1 compound, with a peritectic point of 79 ± 1 °C and that this compound forms a eutectic with *m*-nitrophenol at 67 ± 1 °C at a composition of 72 mol% MNP.

These results are in good agreement with those reported in the literature [63]. Other excellent examples of phase equilibria studies are given by Ford and Timmins [64].

(c) Three-component phase diagrams can be studied by DSC, but care is needed in their interpretation. Thermomicroscopy is most useful in discovering the phases present.

6.3.3 *Liquid crystal studies*

(a) Since there are several transitions on the DSC and different textures on the hot-stage microscope, there are probably liquid crystalline phases. The supercooling of the lowest transition from 92 °C on

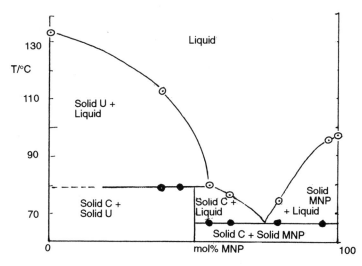

Figure 6.3.2(b) Phase diagram for the urea–*m*-nitrophenol system.

heating to below 75 °C on cooling strongly suggests liquid crystalline behaviour.

(b) The images from the hot-stage microscope show a considerable degree of order. This suggests that the phases are *smectic*. The crystallisation from the melt forms a 'fan-like' texture characteristic of Smectic-A. The transition around 138 °C has a very low ΔH and shows very little optical change. It is probable that the phase below 138 °C is a Smectic-B (hexagonal). Another small DSC peak at 122 °C is accompanied by the 'banding' of the fan texture. This is typical of Smectic-E. The crystalline compound melts to Smectic-E at 92 °C on heating. The phase behaviour is thus:

Crystalline $\leftrightarrows$ Smectic-E $\leftrightarrows$ Smectic-Bhex $\leftrightarrows$ Smectic-A $\leftrightarrows$ Liquid

(c) X-ray diffraction techniques were used to confirm the structures.

6.3.4 *Stability and polymorphism of pharmaceuticals*

(a) Since increasing the temperature will increase reaction rate in accordance with the Arrhenius law, we could heat the drug sample from ambient until decomposition occurs. Hardy [72] points out that if the decomposition is *exothermic* then we could use the ASTM E698 method for evaluating hazardous materials, or use Ozawa kinetics. From the reaction rate equation, it is possible to estimate the time to 10% reacted ($t_{0.1}$) at various temperatures and then to test it by isothermal runs at elevated temperature, for example:

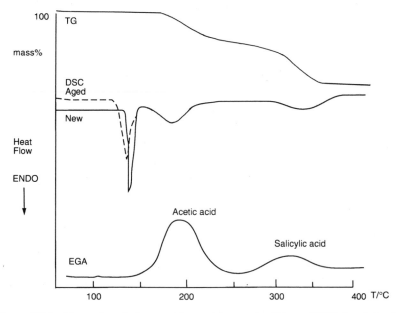

Figure 6.3.4 Thermal analysis curves for aspirin sample: 7.3 mg, 10 K/min, static air (TG, DSC), flowing nitrogen (EGA).

$$\text{Ergocalciferol:} \qquad \ln k = 27.7 - 13930/(T/K)$$
$$\text{Therefore at } 150\,°C, \quad k = 5.46 \times 10^{-3} \ \text{min}^{-1}$$
Assuming first-order kinetics,
$$t_{0.1} = 19.2 \ \text{min}$$

This was close to the value found.

(b) If the sample degrades during the endothermic melting, then the purity will be reduced. This will have two effects: the peak temperature will be lowered and the peak will be broadened or lowered depending on the reaction [76].

A simple experiment with aspirin (acetylsalicylic acid) showed that keeping the sample in air at room temperature changed the peak shape [73]. Running the TG simultaneously with the DSC showed that immediately after the melting at 140 °C, an endothermic decomposition takes place with a mass loss of about 35% (Figure 6.3.4). EGA analysis of the gaseous product identified it (by FTIR) as acetic acid from the reaction to salicylic acid [74].

Polymorphic forms could be detected by DSC or thermomicroscopy [75]. Their interconversion should involve no mass loss on TG.

(c) Loss of solvent would certainly involve a mass loss on TG [73]. Thermomicroscopy often shows water or solvent loss, and this could be identified by a suitable EGA method.

(d) DSC purity experiments need only a few milligrams, and the STA run used 7.4 mg.

6.3.5   *Dynamic mechanical analysis of food products*

(a) The DMA of the bread samples (Figure 6.3.5) clearly shows the changes that occur in samples as they are kept. Three things change:

1. The storage modulus E' measured at $-40\,°C$ decreases from 1 GPa in the fresh sample to a few MPa in the 'overnight' sample.
2. The storage modulus E' measured at $+25\,°C$ increases very slightly from 2 MPa in the fresh sample to 16 MPa in the 'overnight'.
3. tan δ peaks decrease, or disappear!

Possible reasons for these effects are that they are due to the removal of both tightly and loosely bound water. In the fresh sample, the crystalline water increases the subambient modulus and tan δ. The increase in room temperature modulus may be due to the 'plasticising effect' of water on the bread starch.

(b) As a follow-up experiment, many possibilities exist. The reference paper studied pasta material as a function of immersion time.

Other bread samples or other storage conditions could be investigated. If the effects are due to water loss, isothermal TG might be used under the same storage conditions.

6.4.1   *Carbon oxidation*

**6.4
Other materials**

(a) Figure 6.4.1(a) on p. 237 consists of linked straight line segments for each 20-min plot. The rates can be considered as zero order and the rate of reaction calculated from the percent mass loss over a given time.

(b) Figure 6.4.1(c) shows the Arrhenius plot for the example given. The slope is -2.05×10^{-4} and thus the value of E is calculated to be 171 kJ/mol and $A = 1.57 \times 10^4\ s^{-1}$.

(c) A plot of the specific reaction per unit area against the surface area for 20 carbon black samples with areas between 30 and 141 m^2/g shows a general trend for low surface area carbon blacks to have faster rates per unit area than the high surface area carbons. Figure 6.4.1(d) shows the collected data at 773 K, with the rate data normalised to Cabot N774 as unity. There is considerable scatter but the trend is clear at this temperature and also at 793 and 813 K. The full data are given in [82].

For a particular series of reactions there is frequently a relationship

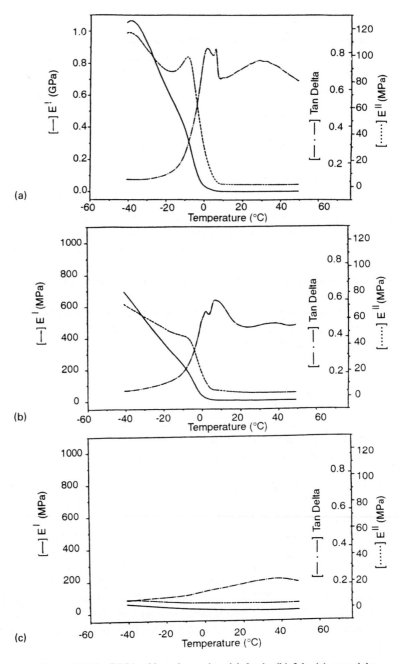

Figure 6.3.5   DMA of bread samples: (a) fresh; (b) 3 h; (c) overnight.

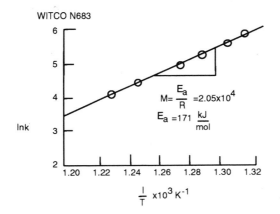

Figure 6.4.1(c)   Arrhenius plot for oxidation of carbon black Witco N683.

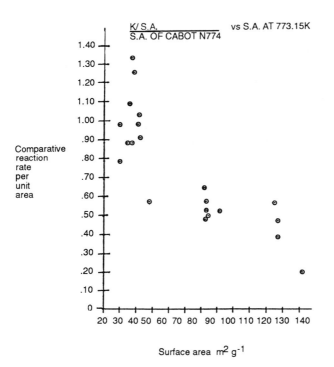

Figure 6.4.1(d)   Plot of specific reaction rate per unit area against the total surface area
for the oxidation of carbon blacks at 773 K.

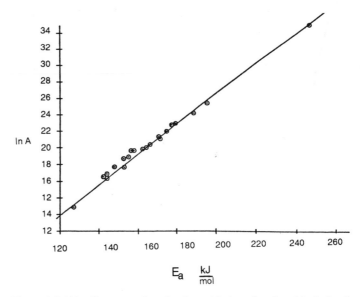

Figure 6.4.1(e)   Compensation plot for oxidation of carbon blacks in air.

between the pre-exponential factor A and the activation energy E, called the 'kinetic compensation effect' [83]. If $\ln(A)$ is plotted against E, a straight line graph is often obtained although the individual data may seem to be rather random. This compensation effect plot is shown as Figure 6.4.1(e).

According to Grisdale [84] and Smith and Polley [85] the rate of oxidation is about 17 times faster in the direction parallel to the basal planes (alo̶   here as 'edge atoms') than perpendicul   ⟩ns show a preponderance of 'edge atoms   ⟩tion leading to a faster rate of oxidation p⟨

This inter   for the reinforcing action of carbon blac   is clearly seen in elastomer– carbon blac   of high surface area, suggest- ing in the li⟨   action between the elastomer and the bas   ⟩h imparts reinforcing action.

6.4.2   *Proximate analysis of coals (Figure 6.4.2(a))*

(a) The loss to 110 °C is 0.70 mg, or 2.8%. This low value for the moisture content is typical for an anthracitic coal.

(b) Loss in nitrogen between 110 and 900 °C is 3.63 mg, or 14.5%, which is the volatiles content.
(c) After switching to oxygen when all the flammable volatiles had gone, the mass decreases by 18.02 mg or 72.1%. This is the fixed carbon.
(d) The residue of 2.65 mg or 10.6%, stable at 900 °C in oxygen, is the ash content.

Values for anthracitic and other coals have been shown to agree well with the ASTM values for the same coals [91(a)].

The calorific value can be estimated by the use of the Goutal equation, or a modification thereof. For anthracitic coals, the equation may be written [91(b)].

$$P = 343(C+V)$$

where C is the percentage fixed carbon on a moisture-free basis and V is the percentage volatile matter on a moisture-free basis.

For the above, $C = 74.1\%$, and $V = 14.9\%$, therefore

$$P = 30\ 527\ J/g = 13\ 133\ BTU/lb$$

Other thermal techniques used for coals include DSC to determine the calorific content. Figure 6.4.2(b) shows the DSC of a coal similar to the one used above, where a small coal sample is run in oxygen and the exothermic peak between 200 and 600 °C measured. The area corresponds to 20.67 J for 0.6881 mg coal, or 30 kJ/g or 12 924 BTU/lb.

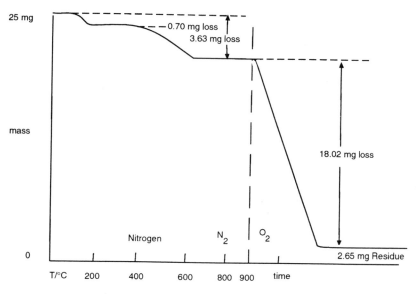

Figure 6.4.2(a)   TG proximate analysis of a coal.

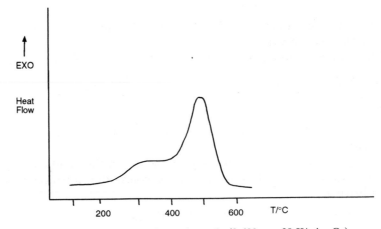

Figure 6.4.2(b)   DSC of a coal sample (0.688 mg, 20 K/min, O_2).

6.4.3   *Oil testing*

(a) The traces clearly show that the onset of oxidation and the onset of discoloration and/or deposit formation occur at the same temperature and time.

   The volatilisation of the oil in an *open* pan is slightly troublesome. If the system could be operated under pressure, or in a crucible constructed for observation under pressure, this could improve baselines.

   The temperature of oxidation onset is 510 K (237 °C), which is in reasonable agreement with other data.

   Comparison of four other oil samples showed a rather poor correlation of the AOM test with both DSC onset time and DSC/RLI onset temperature.

(b) Other techniques used to investigate oxidation of oils, both mineral and edible, include infrared and UV spectrometry, gas chromatography and electron spin resonance (ESR) spectrometry.

   The onset of oxidation often gives a better indication of the start of reaction than, say, IR since the concentration of product species is very low at this point.

6.4.4   *Soil analysis*

(a) The measurement of soil pH, of particle size and mineral microscopy should all be carried out on the original sample as heating will destroy the texture. Removal of organic matter by peroxide oxidation would also be carried out at this stage.

(b) *Sample 1*

Moisture loss (to 150 °C)	18%
Organic matter (150–550 °C)	35%
Mineral content at 600 °C	47%
Mineral breakdown at 750 °C, residue	30%

Sample 2

Moisture loss (to 150 °C)	12%
Organic matter (150–550 °C)	5%
Mineral content at 600 °C	83%
Mineral breakdown (not observed)	

(c) It is not possible to identify the minerals present from thermal analysis alone. X-ray diffraction or geological chemical analysis would be needed.

In Sample 1, the decomposition with a mass loss of some 36% near 800 °C *might* suggest that calcium carbonate is present and this should be investigated. Chemical tests did *not* confirm the presence of carbonates.

Clay contents higher than around 40% cause problems owing to the dehydroxylation of the clays in the same range as the other reactions. Neither sample here showed evidence of high clay content by other tests.

6.4.5 *Catalyst studies*

(a) The formation of methane causes a strong exotherm which reaches a maximum value at around 250 °C due to the almost complete reaction of the carbon monoxide with the hydrogen. With large amounts of catalyst, there may be some 'overheating' due to the thermal lag within the sample.

(b) The presence of catalytic centres on the DSC cell surface will certainly affect the reaction. The cell must be kept clean and the surface area of the catalyst high in comparison with the cell surface.

(c) Kinetic studies on the catalysed reaction could be conducted by varying the CO concentration supplied, and also the hydrogen pressure, or by measuring the partial pressures of the components independently.

6.4.6 *Survival of Antarctic micro-arthropods*

Two events occur during cooling. At around −5 °C, activity slows, but this is hardly detected on the DSC curve. At −14 °C there is freezing exotherm. This is well below the equilibrium freezing point expected, but fairly typical for this species.

The enthalpy change of about 466 J/g may be compared with that of water (at 0 °C) of 2257 J/g. This suggests a water content of about 20%, assuming complete freezing (or only 20% of the body water froze).

References

1. J. Fenerty, P.G. Humphries, J. Pearce, *Proc. 2nd ESTA*, Heyden, London, 1981, p. 293; and *Thermochim. Acta*, 1983, **61**, 319.
2. Z.D. Zivkovic, D.F. Bogosauljevic, V.D. Zlatkovic, *Thermochim. Acta*, 1977, **18**, 235, 310.
3. P.A. Barnes, F.S. Stone, *Thermochim. Acta*, 1972, **4**, 105.
4. D. Dollimore, T.J. Taylor, *Thermochim. Acta*, 1980, **40**, 297.
5. E. L. Charsley, S. B. Warrington, T. T. Griffiths, J. Queay, *Proc 14th Int. Pyrotechnic Seminar, New Jersey*, 1989, p. 763.
6. E. L. Charsley, C. Walker, S. B. Warrington, *J. Thermal. Anal.*, 1993, **40**, 983.
7. G. W. C. Kaye, T. H. Laby *Tables of Physical and Chemical Constants*, 14th edn, Longman, London, 1973.
8. S. Gordon, C. Campbell, *Anal. Chem.*, 1955, **27**, 1102.
9. E L Charsley, S St J Warne, S B Warrington, *Thermochim. Acta*, 1987, **114**, 53.
10. R W Carling, *Thermochim. Acta*, 1983, **60**, 265.
11. A P Gray, P-E Thermal Analysis Application Study 1, 1972.
12. M J O'Neill, *Anal. Chem.*, 1966, **38**, 1331.
13. (a) M.J. Richardson, N.G. Savill, *Thermochim. Acta*, 1979, **30**, 327; (b) M.J. Richardson, *Thermochim. Acta*, 1993, **229**, 1.
14. G.T. Furukawa, T.B. Douglas, R.E. McCoskey, D.C. Ginnings, *J. Res. Nat. Bur. Stand.*, 1956, **57**, 67.
15. F. Jasim, M.M. Barbooti, K.I. Hussain, *Proc 7th ICTA*, Wiley, Chichester, 1982, p. 687; and *Thermochim. Acta*, 1982, **58**, 289.
16. A.A. Zinovev, L.T. Chudinov, *Zh. Neorg. Khim.*, 1956, **1**, 1772.
17. S. Gordon, C. Campbell, *Anal. Chem.*, 1955, **27**, 1102.
18. R.J. Acheson, P.W.M. Jacobs, *Can. J. Chem.*, 1969, **47**, 3031.
19. C80 Brochure: File 6, Sheet 8. SETARAM.
20. S. Gordon, C. Campbell, *Anal. Chem.*, 1955, **27**, 1102.
21. D. Fatu, *Thermochim. Acta*, 1988, **133**, 209.
22. F.R. Bichowsky, F.D. Rossini, *Thermochemistry of Chemical Substances*, Reinhold, New York, 1936.
23. P. Le Parlouer, in *Analytical Calorimetry*, Vol. 5, J.F. Johnson, P.S. Gill (eds), Plenum, New York, 1984, p. 179.
24. J. Brandrup, E.H. Immergut (eds), *Polymer Handbook* (3rd edn), Wiley, New York, 1989.
25. ASTM D3418–82: *Standard Test Method for Transition Temperatures of Polymers by Thermal Analysis*.
26. ASTM D3850–84: *Standard Test Method for Rapid Thermal Degradation of Solid Electrical Insulating Materials by Thermogravimetric Method*.
27. R.B. Cassel, B. Twombly, *Amer. Lab.*, January 1991.
28. J.H. Magill, *Polymer*, 1965, **6**, 367.
29. A. Wlochowicz, M. Eder, *Thermochim. Acta*, 1988, **134**, 133.
30. R.F. Schwenker, R.K. Zuccarello, *J. Polym. Sci. C*, 1964, **6**, 1.
31. J. Brandrup, E.H. Immergut (eds) *Polymer Handbook* (3rd edn), Wiley, New York, 1989.
32. F. Rodriguez, *Principles of Polymer Systems* (2nd edn), McGraw-Hill, Singapore, 1983.
33. B. Cassel, M.S. Feder, M.M. Lebrun, *PIQuality*, 1992 (2).
34. E.M. Barrall II *et al.*, *J. Appl. Polym. Sci.*, 1965, **9**, 3061.
35. J.M. Barton, *Polymer*, 1992, **33**, 1177.
36. K. Horie *et al.*, *J. Polym. Sci. A.*, **1**, 1970; **8**, 1357
37. W.J. Sichina, *Dupont Application Brief*, TA–93.
38. T. Ozawa, *Bull. Chem. Soc. Japan*, 1965, **38**, 1881.
39. K.J. Saunders, *Organic Polymer Chemistry*, Chapman & Hall, London, 1973.
40. E.W. Orrell, R. Burns, *Plastics & Polymers*, 1968, **36**, 469.

41. E.L. Charsley, M.R. Newman, S.B. Warrington, *Proc. 16th NATAS Conf., Washington*, 1987, p. 357.
42. B. Cassell, *Perkin-Elmer Thermal Analysis Application Study, #19, Characterisation of Thermosets*, 1977.
43. V.A. Era *et al.*, *Angewandte Makromoleculare Chemie*, 1976, **50**, 43.
44. (a) ASTM D3350: *Polyethylene Plastics Pipe and Pipe Fittings*; (b) ASTM D3895: *Copper-induced Oxidation Induction Time*.
45. *Mettler Application Notes*, Nos 3002 and 3105.
46. L.M. Moore, G.P. Marshall, N.S. Allen, *Polym. Deg. & Stab.*, 1988, **28**, 1.
47. Netzsch *Onset*, September 1992.
48. ASTM D4092.
49. G.W.C. Kaye, T.H. Laby, *Tables of Physical and Chemical Constants* (14th edn), Longman, London, 1973.
50. F. Rodriguez, *Principles of Polymer Systems* (2nd edn), McGraw-Hill, Singapore, 1983.
51. R.G. Weatherhead, *FRP Technology*, Ch. 9, Applied Science, London, 1980.
52. *Perkin-Elmer Application Example*, PE-TAN 36.
53. M.G. Lofthouse, P. Burroughs, *J. Thermal Anal.*, 1978, **13**, 439.
54. R.E. Wetton, *Developments in Polymer Characterisation*, Ch. 5, Applied Science, London, 1986.
55. F. Rodriguez, *Principles of Polymer Systems* (2nd edn), McGraw-Hill, Singapore, 1983, 329.
56. R.E. Wetton *et al.*, *Amer. Lab.*, 1986 (January), 70.
57. T.A. Instruments, *Application Note*, TS-6.
58. ASTM E928–83, Philadelphia.
59. E.F. Palermo, J. Chiu, *Themochim. Acta*, 1976, **14**, 1.
60. G. Becket, S.B. Quah, J.O. Hill, *J. Thermal Anal.*, 1993, **40**, 537.
61. B. Joseph, *J. Thermal Anal.*, 1993, **40**, 1447.
62. F.D. Ferguson, T.K. Jones, *The Phase Rule*, Butterworth, London, 1966, p. 51.
63. *Landolt–Bornstein Tables*, Band II, Vol. 2c, II, pp. 258–9.
64. J.L. Ford, P. Timmins *Pharmaceutical Thermal Analysis*, Ellis Horwood, Chichester, 1989.
65. J.M. Gines *et al.*, *J. Thermal Anal.*, 1993, **40**, 453.
66. J.L. Jorda *et al.*, *J. Thermal Anal.*, 1988, **34**, 551.
67. G.W. Gray, *Molecular Structure and Properties of Liquid Crystals*, Academic Press, London, 1962.
68. W.P. Brennan, A.P. Gray, *Perkin-Elmer Thermal Analysis Applications Study*, #13, 1974.
69. J.V. Wood, *Thermal Analysis of Electronic Materials*, T.A. Instruments (DuPont) Publication, 1980.
70. H.G. Wiedemann, *Mettler Application Brief*, No. 805.
71. J.W. Brown, D. Hurst, J. O'Donovan (unpublished work).
72. M.J. Hardy, *Proc. 7th ICTA*, Wiley, Chichester, 1982, pp. 876, 887.
73. J. Joannou, Rheometric Scientific Ltd. Private communication.
74. TA Instruments, *TA Hotline*, 1993, **1**, 8.
75. J. Masse, R. Malaviolle, A. Chauvet, *J. Thermal Anal.*, 1979, **16**, 123, 343.
76. J. Lubkowski *et al.*, *Thermochim. Acta*, 1989, **155**, 29.
77. C.G. Biliaderis *et al.*, *J. Food Sci.*, 1980, **45**, 1669.
78. M. Wooton, A. Bamunuarachi, *Starke*, 1979, **32**, 262.
79. W.M. Jackson, J.F. Brandts, *Biochem. J.*, 1970, **9**, 2293.
80. J. Foreman, *TA Hotline*, 1991, **4**, 6.
81. P. Roulet *et al.*, *Food Hydrocolloids*, 1988, **2**(5), 381.
82. J.A. Azizi, D. Dollimore, P.J. Dollimore, G.R. Heal, P. Manley, W.A. Kneller, W.J. Yong, *J. Thermal Anal.*, 1993, **40**, 831.
83. A.K. Galwey, *Adv. Catal.*, 1977, **26**, 247.
84. R.O. Grisdale, *J. Appl. Phys.*, 1956, **24**, 1288.
85. W.R. Smith, M.H. Polly, *J. Phys. Chem.*, 1956, **60**, 689.
86. M. Odlyha, A. Burmester, *J. Thermal Anal.*, 1988, **33**, 1041.
87. R.T. Morrison, R.N. Boyd, *Organic Chemistry* (2nd edn), 1966, p. 684.
88. M. Odlyha, *Thermochim. Acta*, 1988, **134**, 85.
89. F. Preusser, *J. Thermal Anal.*, 1979, **16**, 277.

90. ASTME 1131, Compositional analysis by thermogravimetry.
91. (a) R.L. Fyans, *Perkin-Elmer Application Study*, #21; (b) C.M. Earnest, R.L. Fyans, *Perkin-Elmer Application Study*, #32.
92. R.L. Hassel, *Dupont Application Brief*, TA-54.
93. J.W. Cumming, *Thermochim. Acta*, 1989, **155**, 151.
94. R. L. Blaine, *Amer. Lab.*, 1974 (Jan.).
95. J.A. Walker, W. Tang, Society of Automotive Engineers, *SAE Technical Paper*, #801383, 1980.
96. K. Rajeshwar, D.B. Jones, J.B. DuBow, *Anal. Chem.*, 1981, **53**, 121.
97. A. Millington, D. Price, R. Hughes, *J. Thermal Anal.*, 1993, **40**, 225.
98. B.C. Burros, *Int. Lab.*, 1986 (April), 18.
99. R.L. Hassel, *J. Amer. Oil Chem. Soc.*, 1976, **53**, 179.
100. F.N. Zeria, R. A. Moore, Society of Automotive Engineers, *SAE Technical Paper*, #890239, 1989.
101. R. Brewer, *Fabric and Mineral Analysis of Soils*, Wiley, Chichester, 1969.
102. M. Schnitzer, J.R. Wright, I. Hoffman, *Anal. Chem.*, 1959, **31**, 440.
103. R.C. Mackenzie, *Proc. 7th ICTA*, Wiley, Chichester, 1982, p. 25.
104. T. Ishii, *Thermal Analysis; Comparative Studies on Materials*, H Kambe, P D Garn (eds), Halsted Press, 1974, p. 175.
105. J.J. Birmann, H. Den Daas, F.J.J.G. Janssen, *Thermochim. Acta*, 1988, **133**, 169.
106. M Malinowski, S Krzyzanowski, *J. Thermal Anal.*, 1972, **4**, 281; and *Proc. 1st ESTA*, Heyden, London, 1976, p. 128.
107. M.A. Vanice, *J. Catal.*, 1976, **44**, 152.
108. G. Hakvoort, L.L. van Reijen, *Proc. 7th ICTA*, Wiley, Chichester, 1982, p. 1175.
109. G. Munteanu, E. Segal, A. Butucelea, *Thermochim. Acta*, 1988, **133**, 137.
110. M.R. Worland, W. Block, *J. Insect Physiol.*, 1986, **32**, 579.
110a. W. Block, *Phil. Trans. R. Soc. Lond., B*, 1990, **326**, 613.
111. T.N. Hansen, J.G. Baust, *Biochim. Biophys. Acta*, 1988, **957**, 217.
112. W. Block, *Acta Oecologica*, 1994, **15** (in press).

Appendix
Solutions to problems in
Chapters 1–5

1. Substance is low melting, and probably covalent. It burns in air with no residue, so is possibly an organic (C,H,O) compound. The melting could be studied by DSC or TMA under nitrogen to prevent burning. The burning could be studied by TG in air. (*Example*: naphthalene.)
2. The colour change, residue and brown fumes suggest a hydrated transition metal nitrate. (*Example*: cobalt nitrate hydrate.)
3. Thermodilatometry. This is a special case of thermomechanical analysis.
4. 'Thermoelectrometry is a family of techniques in which an *electrical* property is measured against *time* or temperature while the *temperature* of the sample in a specified *atmosphere* is programmed.'
5. (a) The sample formula or source is not specified.
   (b) The apparatus (DSC, TG, etc.) is not specified.
   (c) The rate, atmosphere and mass of sample are not given.
6. Thermomicroscopy could be used to study the changes and evolved gas analysis to study the gaseous products. The solid products could be analysed by X-ray diffraction or wet chemical methods.

1. Moisture loss, sublimation and gas adsorption all involve a change of mass. Melting does not, and is not detected by TG.
2. (a) Would be detected since the gases are not weighed, but the solid increases in mass; (b) would be detected, since CO_2 gas is lost; (c) would *not* be detected, since the double decomposition involves no mass change.
3. As always, remember SCRAM!

   S check the provenance, formula and state of the sample
   C select an inert crucible, suitable to the temperatures to be used
   R select a heating rate and start and end temperatures
   A set the purge gas and flow rate
   M decide what mass of sample you wish to run

For stability testing, you may have to perform several runs under different conditions, e.g. in air and nitrogen.

4. Carbonates usually react with acids to give off carbon dioxide and yield a salt:

$$Mg\ CO_3 + H_2C_2O_4(aq) = MgC_2O_4 \cdot xH_2O + CO_2$$

A

The loss 2.23 mg (24.2%) to 220 °C could be of the hydrate water. The RMM of the compound **A** is $(112 + x \times 18)$. If $x = 2$ then the loss would be $[2 \times 18/148] \times 100 = 24.3\%$

$$MgC_2O_4 \cdot 2H_2O = MgC_2O_4 + 2H_2O$$

As we have seen with calcium oxalate, the decomposition may go through the carbonate to the oxide. Since there is only one stage here, it is probable that the oxide is formed directly, since the residue is 27.0%, compared with the calculated residue of 27.0% for the total loss.

The second reaction must be:

$$MgC_2O_4 = MgO + CO + CO_2$$

5. The final residue must be silver metal plus copper oxide, totalling 48% and with a loss from the previous stage, copper oxide plus silver nitrate, of 67%. Thus, a loss of 19% for loss of 'NO$_3$', RMM 62, in the reaction:

$$AgNO_3 = Ag + NO_2 + \tfrac{1}{2}O_2$$

corresponds to a silver (RAM 108) content of 33.1%. Since the AgNO$_3$ (RMM 170) would then be 52.1%, the CuO must be

$$(48.0 - 33.1) = (67.0 - 52.1) = 14.9\%$$

The copper (RAM 63.5) must be 11.6% of the original *salts*, or the original alloy has

$$100 \times [33.1/(33.1+11.6)] = 74.0\%\ \text{silver}$$

6. The initial loss is chiefly due to the breakdown of the poly(vinyl acetate) with loss of ethanoic acid (acetic acid). A loss of 15% due to ethanoic acid (CH$_3$COOH, RMM 60) means that the vinyl acetate moiety,

$$—[—CH_2—CH(OOC \cdot CH_3)–]—, \quad \text{RMM } 86,$$

must make up 21.5% of the copolymer.

7. Suggest to your colleague that his or her procedure should be more rigorously standardised! (SCRAM!) Are similar masses of sample being used each time? Are the conditions the same? If the heating

rates are different, temperatures will be different too. If all rules have been obeyed, it is probable that your colleague's sample is *not* pure, or that it is not homogeneous. This actually happened with samples of magnesium hydroxide that were damp!

Chapter 3
Differential thermal analysis and differential scanning calorimetry

1. (a) Loss of moisture is endothermic and would be detected.
   (b) Sublimation is endothermic.
   (c) Desorption is endothermic.
   (d) Polymer softening, unless at a glass transition, may not be detected since it is a mechanical, more than a thermal transition. DMA would detect it.
2. As always: SCRAM! In this case the heating rate will have the effect of changing the appearance of all the observed events, since:
   (a) $\Delta y = \beta \cdot C_p$, the baseline shift will double each time the heating rate is doubled;
   (b) $\Delta H = \int \Delta P \, dt = K \int \Delta T \, dt$, the integration must be conducted with respect to **time**. If we change the time by heating more rapidly, we shall change the area.
3. This system forms a continuous series of solid solutions.
4. The fractions melted may be plotted versus temperature, remembering that T should be in Kelvin!

T/K	$1/F$	
407.8	0.00	The slope of the plot $= -0.32$
406.7	3.45	$= -(RT_0^2/\Delta H)\cdot x_B$
406.2	5.00	$= -(8.314 \times 407.8^2/53\ 300)\cdot x_B$

Thus $x_B = 0.0123$, or the purity of the phenacetin is 98.77%.
5. $\Delta H = -157\ -297 - 0 - (-800) = 346$ kJ/mol. This indicates an endothermic reaction, as found. This is an estimate at 298 K, but the reaction actually occurs around 1000 K. Data on the heat capacities would be needed to correct the value.
6. The first run shows the T_g of the initial material. The second shows the curing reaction and the third only the T_g of the cured sample.
7. Peaks due to mass loss occur only on first heating whereas inversions usually appear 'inverted' on cooling curves. If cooling is not possible, a second heating should reproduce an inversion but not a peak due to loss of mass.
8. A mineral of 'ideal' composition is quite unusual. Mixtures of kaolin (K) and felspar (F) are made by weighing out various proportions of K and F and mixing them well. Each mixture, and the suspect material, is subjected to both DTA and TG runs. The mass loss at 600 °C is noted, and a graph of mass loss versus kaolin content is plotted. The %K of the suspect material may then be read from the graph. The height or area of the peak at 600 °C may also be correlated with %K.

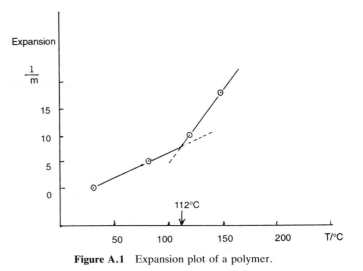

Figure A.1 Expansion plot of a polymer.

Chapter 4
Thermomechanical,
dynamic mechanical
and associated methods

1. A simple system might involve a light source and photocell to sense the movement of a shade across a light path due to the expansion of the sample. Disadvantages could arise from optical interference, heat distortion, or the non-linear response of the detection system.

    Such systems are described in B.K. Jones, *Electronics for Experimentation and Research*, Ch. 8, Prentice Hall, London, 1986.

2. From the laws of expansion:

$$\Delta l = \alpha \cdot l \cdot \Delta T = \alpha \times 3.00 \times 10^{-3} \times 100$$
$$\alpha = 2.5 \times 10^{-5} \ K^{-1}$$

   This is approximately the value for aluminium. The volume expansion may be calculated from:

$$V_T = V_0(1 + \alpha \cdot \Delta T)^3 = 150(1 + 2.5 \times 10^{-5} \times 100)^3$$
$$V_T = 151 \cdot 128 \ cm^3$$

   The coefficient of volume expansion is then

$$(V_T - V_0)/(V_0 \cdot \Delta T) = 7.52 \times 10^{-5} \ K^{-1}$$

   or approximately 3α.

3. A simple plot of Δl versus T gives two lines intersecting at about 112 °C. Note that with this small amount of data, the full characterisation of the glass transition is not possible (Figure A.1).

4. Substituting the values in the given equation:

$$F = 0.020 \times 9.81 = 0.1962 \ N$$
$$R = 2.0 \times 10^{-3} \ m, \ d = 0.113 \times 10^{-3} \ m$$

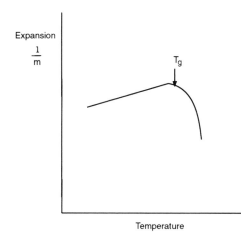

Figure A.2   Penetration TMA for rubber sample.

Thus $E = 2.05 \times 10^6$ Pa. This compares well with the Young's modulus value of 2.3×10^6 Pa.

The TMA curves might have the appearance of Figure A.2.

5. Normally for thin film or fibre samples, the most appropriate mode of operation is in *tension*. The first difficulty is that of avoiding *buckling* the sample. This is achieved by applying, in addition to the dynamic force, a constant tension sufficiently large to prevent the total force becoming compressive, but not so large as to cause excessive creep! Typically it is 1.1 to 2 times the dynamic force. An additional problem is encountered when the modulus of the sample changes. As this happens, the force required to maintain a constant dynamic strain changes. During a glass transition, this may change by three orders of magnitude, so the constant tension must change too. This can be done by holding the *ratio* of the dynamic and constant forces at the same value throughout the experiment. The variation of these forces is shown in Figure A.3.

6. If the sample prepared is too brittle to be clamped in the normal way as *supported* technique may be used. A convenient technique is to impregnate a glass fibre braid with the solution and then allow it to dry before mounting the braid in the DMA. This relates to 'torsional braid analysis', TBA. One disadvantage of this approach is that it is not possible to measure the *absolute* value of the polymer modulus. Another is that the sample may interact strongly with the glass fibre braid which could distort the results by, for example, increasing the apparent T_g or by the enrichment of a component of the sample at the sample–support interface.

In this case the sample is a mixture. Experiments must be performed on each component to establish their T_g values. If they are different

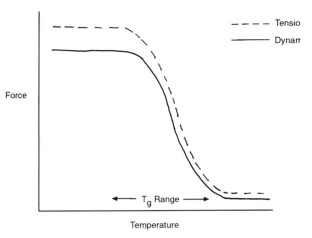

Figure A.3   Forces during operation with a thin film or fibre.

then the results for the mixture provide clues for whether the two polymers are miscible. If there are two peaks in the tan δ plot that have the same values as the pure component polymers, then they are completely immiscible. If a single T_g is seen at an intermediate value, then they are miscible.

7. Dynamic mechanical analysers are rarely adapted for making measurements on liquids as they require that the sample be clamped into position. A convenient way of investigating cross-linking behaviour is to mount a glass fibre braid in the DMA (as in Problem 6) and then impregnate it with a known volume of liquid sample. If heating is required to initiate cross-linking then an appropriate heating programme is used.

The storage modulus will begin to rise when the gel point is reached, also marked by a peak in tan δ. As cross-linking proceeds the T_g of the sample will increase and may rise to meet the cure temperature. When this happens the sample will vitrify. This is also marked by an increase in storage modulus and a peak in tan δ. The cure cycle can be said to be complete when the temperature programme, which may be imitating the environment in a real industrial process, has ended, or when the storage modulus reaches a plateau. After this stage, the T_g of the sample and any residual cure can be measured by cooling the sample and performing a temperature scan in the normal way. After the T_g the modulus may rise if the cure reaction is pushed further.

The *cross-link density* of the cured material can be found by measuring its modulus above the T_g. This cannot be done on a support, as the modulus of the polymer–braid composite is difficult to relate to the modulus of the polymer. Consquently, a cured sample must be prepared before running in the DMA. Different frequencies should be tried to gauge the effect of this variable, but a plateau value should be reached at low frequencies equivalent to the equilibrium storage modulus. The cross-link density may then be calculated using the equation in the text.

8. If the two peaks arose from a single sample, then the upper temperature transition, involving such a large drop in modulus, would be assigned as a *glass transition* (or 'alpha transition') caused by large-scale cooperative motions that involve the main polymer chain. The lower temperature transition would then be assigned as a *beta-transition*. Such secondary transitions are caused by localised motions which involve little motion of the main polymer chains. These include side-chain motions, or in-chain events such as rotations and large angle torsional librations.

   If the sample were composed of *two components* then the upper temperature event would still be assigned as a glass transition, but the lower temperature event might now be due to a second phase that has a lower T_g. Whether the lower temperature event is due to a secondary transition or a second T_g may be investigated by measuring the relaxation kinetics as a function of temperature. If it follows a WLF expression, then it should be a T_g, whereas a secondary transition normally follows an Arrhenius equation.

9. (a) The value of $\log(a)$ may be compared with the calculated value from the given WLF constants:

$T/°C$	$\log(a)$ [obs.]	$\log(a)$ [WLF]
135	3.36	3.36(2)
140	1.57	1.56(6)
145	0	0
150	−1.38	−1.37(8)
155	−2.60	−2.59(9)

   We could also perhaps show the agreement with the Arrhenius expression. If the temperature of maximum tan δ for 1 Hz is taken as 145 °C (418 K), then the maxima occur at other temperatures when the frequency is shifted, e.g. using a frequency of 398 Hz, the maximum will be at 155 °C (428 K). A plot of ln (f) against $1/T_{td}$ gives a smooth curve. This suggests that the transition studied may be an alpha transition.

   (b) The lowest frequency will be 10^{-2} Hz lowered by $10^{-2.6}$ Hz, that is $10^{-4.6}$ Hz.

10. Using the dimensions given, the fraction of polymer in the final solvent-swollen material will be:

$$v_2 = (1.65)^3/(2.83)^3 = 0.198$$

With the given data:

$$n = \frac{-[\ln(1 - 0.198) + 0.198 + 0.391 \, (0.198)^2]}{106.3[(0.198)^{1/3} - (0.198)/2]}$$

$$n = 1.469 \times 10^{-4} \text{ mol/cm}^3$$

Chapter 5
Simultaneous
techniques and product
analysis

1. With modern instrumentation, even very small samples may be analysed quite easily. As a first attempt, we should choose *non-destructive methods*, such as IR or microscopy. If thermal data are needed, DSC and thermomicroscopy should provide information on polymorphism, melting, purity and thermal stability. With a very small sample, *simultaneous DSC–thermomicroscopy* and/or *DSC–FTIR* could be very informative.

2. If the appropriate traces in Chapter 2 and Chapter 3 are studied, it should be clear that in normal circumstances, the detection will be as follows:

Event	DTA	TG
(a) glass transition	Yes	No
(b) plasticiser loss	Yes	Yes
(c) residual curing	Yes	Possibly
(d) crystallisation	Yes	No
(e) melting	Yes	No
(f) oxidation	Yes	Yes
(g) degradation	Yes	Yes

3. The evolution of gases may be resolved by temperature or time or by spectral dispersion.
   (a) H_2O from (i), (ii) at about 100 °C and from (ii), (iv) between 200 and 700 °C.
   (b) CO_2 from (iii) around 800 °C and from (iv) between 200 and 600 °C.
   (c) CO from (iv) between 200 and 500 °C.
   (d) SO_2 from oxidation of (v).
   [Reference: D.J. Morgan, S.B. Warrington, S. St J. Warne, *Thermochim. Acta*, 1988, **135**, 207.]

4. The reaction at about 100 °C might correspond to loss of hydrate water, which would give a mass spectrometric peak at $m/z = 18$, as found.

   Peaks due to ammonia ($m/z = 17$) and nitrogen ($m/z = 28$) or carbon monoxide (also $m/z = 28$) occurring at 350 °C suggest decomposition of the anhydrous material, with some loss of CO_2 and water.

   The final loss at 850 °C is the decomposition of the calcium carbonate to calcium oxide with loss of CO_2 ($m/z = 44$). The peak at $m/z = 28$ could be due to the fragmentation to CO or to chemical reduction of the dioxide. We may write the equations:

At about 100 °C:

$$Ca(N_2H_3COO)_2 \cdot H_2O = Ca(N_2H_3COO)_2 + H_2O$$

with a loss of 8.4% against a calculated loss of 8.6%.
At about 300 °C:

$$Ca(N_2H_3COO)_2 = CaCO_3 + 2NH_3 + N_2 + CO$$

with a loss of 42.0% against a calculated loss of 43.3%.
At about 800 °C:

$$CaCO_3 = CaO + CO_2$$

with a further loss of 21.8% against a calculated loss of 21.1%. Chemical and X-ray analyses confirmed the products, but some unidentified material was present.
[Reference: B. Novosel, J. Macek, V. Ivancevic, *J. Thermal Anal.*, 1993, **40**, 427.]

5. Obviously we should know a great deal more about the sample, its history and composition before we can give proper answers, but the evidence presented *suggests*:

   (a) The colour suggests a transition metal salt or complex, possibly containing chromium, copper or cobalt or iron. A colour change at 55 °C suggests a phase change or a decomposition. TG would help here.

   (b) At 110 °C the crystals became opaque. This is frequently associated with hydrate or complex decomposition. The TG should show a mass loss if this is correct. Such a decomposition would be endothermic.

   (c) A colour change to brown might indicate a further decomposition (to oxide?). This could well be endothermic, and would be shown by TG.

   Additional information is required here!

6. Tristearin was noted in Chapter 3 to exhibit polymorphism. The low-temperature α form melts at about 56 °C, and this would decrease the light transmitted through crossed polars. Recrystallisation to form β restores the light intensity, since the crystals will depolarise the light. Melting of the β form at 75 °C takes the intensity back to zero, characteristic of an isotropic liquid.

7. (a) The mass losses correspond closely to the suggested reactions, but we *must* note that all intermediates and products should be checked by independent analytical methods.

   For (i), the water evolution could be checked by EGA, either MS or IR or a specific detector. Similarly the evolution of CO and CO_2 simultaneously in (ii).

   The nature of the solids in all cases should be checked by X-ray diffraction and/or chemical analysis. All these checks were carried out in the reference below.

[Reference: D. Dollimore, D.L. Griffiths, D.Nicholson, *J. Chem. Soc.*, 1963, 2617.]

(b) The original gel is clearly amorphous and has no regular crystalline structure. On heating, it either crystallises (or possibly reacts) to form a crystalline compound.

[Reference: J.A. Navio *et al.*, *J. Thermal Anal.*, 1993, **40**, 1095.]

Index